Y0-BTD-339

5·20·77

The Cost of Electricity

cheap power vs. a clean environment

Gulf Publishing Company
Book Division
Houston, Texas

THE COST OF ELECTRICITY

cheap power vs. a clean environment

Russell G. Thompson
James A. Calloway
Lillian A. Nawalanic

Contributors

L. Ted Moore
Rodrigo J. Lievano
H. Peyton Young

The Cost of Electricity
Cheap Power vs. A Clean Environment

Library of Congress
Catalog Card Number
76-40869
ISBN 0-87201-156-9

Disclaimer

This material was prepared with the support of National Science Foundation Grant No. GI-34459-2, "National Economic Models of Industrial Water Use and Waste Treatment," Russell G. Thompson, Principal Investigator.

Any opinions, findings, conclusions, or recommendations expressed in this publication are those of the authors and do not necessarily reflect the views of the National Science Foundation or the University of Houston.

Principal Investigator's Foreword

The skyrocketing cost of electricity is of great concern to residential and industrial consumers; the public wants to know why these costs are increasing so rapidly, if this inflationary price rise will continue, and what steps can be taken to moderate these spiralling costs. At the same time, the public wants readily available electricity, a cleaner environment, cheap natural gas, and low-cost petroleum fuels. But the scarcity of domestic petroleum, natural gas, and low-sulfur coal is becoming increasingly acute, and using these scarce resources to satisfy one need means they will not be available to satisfy another. Moreover, due to the expanding demand for electricity, the contribution from hydroelectric power will steadily diminish; as for nuclear power, the scarcity of fuel (uranium) is more acute here than in fossil-fueled power generation, and the stalemate between advocates of nuclear power and environmentalists has no foreseeable resolution. Clearly, tradeoffs must be made between a clean environment, independence from foreign energy, and an affordable cost of living.

In fossil-fueled electric power generation, a vast set of interrelationships exist. For example, to maintain a clean environment, strict effluent standards will be necessary, and these in turn will increase the already acute demand for scarce *clean* fuels. With additional demands for these scarce clean fuels, the fuel cost component of the cost of living will rise, and additional imports of expensive foreign crude oil will be needed. This is particularly true if electric utilities are allowed to bid freely for clean natural gas. Because of the incentives to avoid costly stack gas scrubbers and to gain from use of highly efficient combined-cycle generating technology, electric utilities can bid unusually high prices for natural gas; they can effectively bid this gas away from higher-priority users in the residential, commercial, and industrial sectors. Strict environmental standards must be accompanied by appropriate complementary fuel use restrictions if natural gas is to be available at reasonable prices to high-priority users. Thus, we can only have a clean environment at some expense to the goals of energy independence, utility regulation, and electricity costs.

In this study a concerted effort was made to develop an economic framework to show the public the real costs of the interrelated decisions to either control waste discharges to the water and the air or to force electric utilities to burn cheaper, dirty coal instead of clean natural gas and oil products. Also, the real costs of variations in electricity use, low-sulfur coal prices, and capital availabilities are analyzed. This economic framework synthesizes the relevant technical information into a computer-based model of the industry to identify key decision variables and to measure the resource, environmental, and economic consequences of changes in these variables. Historical data are not sufficient to show the economic consequences of major government policy decisions in these areas because past trends have no bearing on the future in a period of such drastic change, when so many variables are involved.

Results of the study show how restrictive waste discharge standards, decreased availabilities of clean fuels, and increased scarcity of capital for utility investment will affect the technical generation structure of the electric power industry. In going from lax to strict environmental standards, the production mix changes from large investments in new high-sulfur coal-burning plants to heavy investments in coal-gasification combined-cycle plants (using high-sulfur coal). The costs of cleaning up air emissions are found to be much greater than the costs of cleaning up wastewater discharges. Strict air emission standards affect the technical structure of generation, but strict wastewater standards affect only the water use and wastewater treatment system. Decreased availabilities of clean fuels greatly increased the costs of environmental clean-up.

Limited capital availabilities and expanded electricity requirements increase clean air and clean water costs. Development of the low-sulfur coal market will depend heavily on the availability of clean natural gas for use in electricity generation. Investments in new coal-burning generation technologies will be affected significantly by the availability and cost of capital.

The research work of this monograph was supported at the University of Houston by a $1,077,000 grant, "National Economic Models of Industrial Water Use and Waste Treatment," from the National Science Foundation, Research Applied to National Needs (RANN) Program. Dr. Larry W. Tombaugh was the Project Monitor from June 1972 to December 1974; Mr. Gordon Jacobs succeeded Dr. Tombaugh as Project Monitor from January 1975 to the termination of the Project in December 1975. These capable Project Monitors provided leadership, support, and encouragement in the fruitful completion of the Project work.

As Principal Investigator, I am indebted to the support of the University of Houston, particularly the College of Business Administration, in providing facilities for completing the monograph and in encouraging graduate student involvement in the Project. I am proud of and grateful to a hardworking loyal Project staff for imaginative, unstinting, and sound contributions. Project staff members have been given authorship recognition for the chapters to which they contributed. Acknowledgement is also gratefully extended to Koichi Inoue, Gordon Otto, Ernest Henley, and Iori Hashimoto.

The dedicated secretarial services of Mrs. Norma Stout Burkhardt, Mrs. Paula Mayfield Hoepner, and Miss Kathy Chan are happily acknowledged; the manuscript went through several "rough" drafts—all of which they typed with patience and good will.

Russell G. Thompson
Principal Investigator
NSF (RANN) Grant GI-34459-2

Contents

The Cost of Electricity

cheap power vs. a clean environment

Introduction

Government policymakers need to know how cost-conscious managers in electric utilities will respond to major changes in government policy *before* policy is changed. Managers of electric utilities respond to policy changes by selecting alternative fuel mixes, generation methods, treatment methods, and investment strategies. Policymakers need to know (1) how much increasingly restrictive waste discharge standards for air and water pollutants will increase the fuel use, plant capital costs, and the cost of generating electricity; (2) how much increasingly restrictive wastewater discharge standards will increase the cost of controlling air pollution, and vice versa; (3) how much the higher prices of crude oil, natural gas, and coals will increase the costs of generating electricity and waste discharge control; and (4) how much capital investment will be needed to modify old plants and to build new ones to meet both the projected growth in electric power consumption and the government's projected environmental goals.

This response information for fossil-fueled electric power generation will be of top priority in the next ten years. Electric power generation consumed 42% of the fossil energy used by all industries (including electric power) in the nation in 1972. FEA predicts that electric power generation will consume 48% of the fossil fuel used by all industries in 1985 (business as usual without conservation, $7 oil, *Project Independence Report*, 1974).[1]

An economic policy model of electric power generation has been developed to answer questions concerning (1) the generation, fuel transformation and waste treatment process changes which will occur with various levels of waste discharge control, fuel availabilities, and capital limitation; (2) the increases in generation cost and fuel use resulting from zero discharge of pollutants to the water and from control of major pollutants; (3) the increased value of clean fuel resulting from implementation of stringent emission controls and varying levels of capital availability; and (4) the demand for additional investment capital by utilities as a function of effluent standards, fuel availabilities, and electricity requirements.

The economic model of electric power generation includes models of both old and new plants of traditional design; these plants may use low-, medium-, or high-sulfur coal, low- or high-sulfur oil, or natural gas. New plants with coal gasification combined-cycle technology are also included in the model; these plants may use low-, medium- or high-sulfur coal to supply the gasifier. All technologies modeled are either presently developed and available for investment or could be operational by 1985. The important ways of substituting processes in power production, fuel use, water use, wastewater treatment, air emission control, and solid waste and brine management are identified and their economics evaluated. Such substitution possibilities are limited by physical laws, electricity requirements, fuel and water availabilities, waste discharge restrictions, and capital availabilities. These limitations have been included in the model. An additional feature of the model is that it treats capital as a resource necessary for the operation of new equipment.

Methods Of Analysis

A linear programming model of electric power production was developed to evaluate the important substitution possibilities in power production, fuel use, input water treatment, water use, investment capital use, wastewater treatment, air emission control, and solid waste management. The fundamental components of the model are illustrated in Figure 1.

The inputs to the model are fuel, water, and capital. The types of fuels considered are oil, coal, and natural gas. These include two grades of oil (low-sulfur and high-sulfur) and three grades of coal (low-, medium-, and high-sulfur). Due to the flexibility of the model, each of the inputs can be varied in price. In the case of water, the price of withdrawal can be varied; in the case of capital, varying availabilities can be considered. Further, different availabilities of each input can be specified.

The power-generating process technologies considered in the model include both old and new fossil fuel-fired plants. Nuclear, peaking, and hydroelectric generation, along with transportation and distribution costs, were specified outside of the model; however, they are included in an expansion of the model for the capital analysis in Chapter 9. Old power plants are assumed to include particulate control units but not sulfur control units; however, the model allows for investments in equipment for the control of sulfur oxides. It is assumed that new plants can include sulfur and particulate control units and that both old and new plants can exclude the use of pollution control equipment in the event of relaxed pollution control standards.

The outputs of the model are electric power and waste discharges to the environment. The latter include SO_2 and particulate emissions to the air; heat, dissolved solids, and suspended solids to the water; and solid wastes to the land. The model may be extended to include analysis of other pollutants such as NO_x to the air and chlorine in the cooling tower blowdown. Consumer demand for electricity is varied as a function of price and personal income; different contributions of nuclear power are considered. Further, the effects of different possible restrictive standards and effluent taxes on amounts of waste effluents are evaluated. The model effectively shows how the various inputs and outputs, their prices, and their availabilities interact with each other, indicating which production and waste treatment configurations are feasible and which are not.

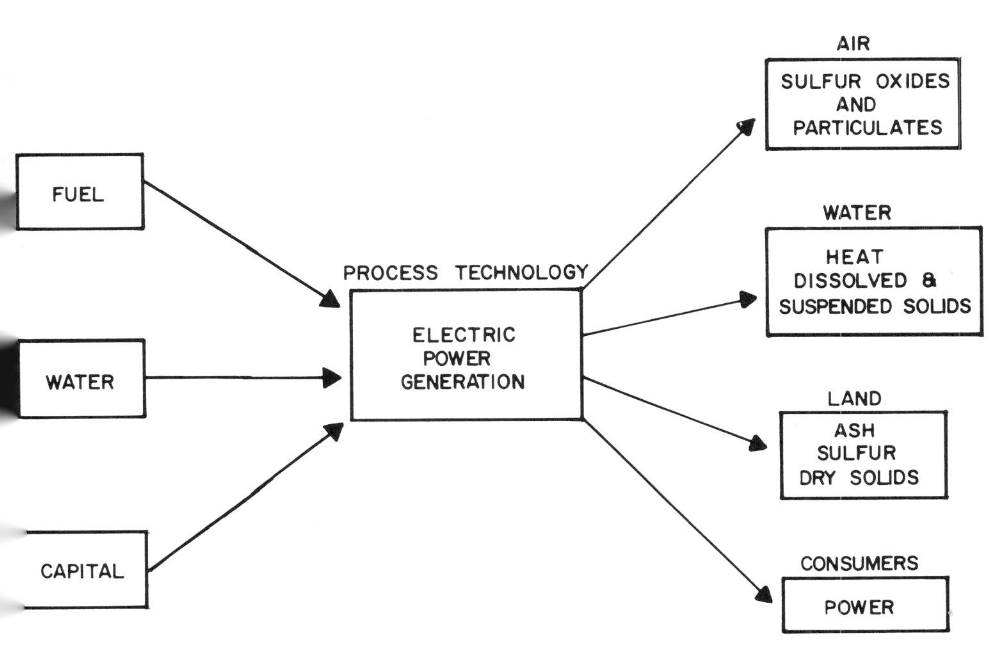

Figure 1. Input-output structure of the electric power generation model.

The electric power model was developed in the following way:

1. Engineering flowgraphs were developed for the power production, fuel use, input water treatment, water use, wastewater treatment, air emission control, and solid waste and brine management systems.
2. The important ways of substituting processes in power production, fuel use, input water treatment, water use, wastewater treatment, air emission control, and solid waste and brine management were identified from the flowgraphs.
3. For each substitution possibility, the quantity of each input and the quantity of each output are specified in a linear model with the input-output coefficients organized in an array of numbers called a tableau; the costs of different operating levels of the major generation and treatment units are calibrated using nonlinear engineering submodels.
4. The substitution possibilities are limited by physical laws governing water and fuel use in the modeled plant; the substitution possibilities are further limited by restrictions on fuel and water availabilities, limitations on waste discharges and capital availabilities, and requirements to produce specified quantities of electricity.

5. Cost-minimizing solutions to the linear model are calculated for each specification of effluent standards for waste discharges, availabilities of fuel, water, and capital, and requirements for electricity; cost-minimizing solutions may also be calculated for different locations.
6. Each least-cost solution gives (1) the average cost of producing a kilowatt hour of electricity at each requirement level; (2) the marginal cost of producing a kilowatt hour of electricity with each generation mix; (3) the use of each type of fuel; (4) the marginal value of an additional unit of each scarce fuel; (5) the amount of water withdrawn and the amount used for in-plant consumption and the amounts of waste discharges; (6) the marginal cost of meeting each waste discharge restriction; (7) the investments (in constant 1973 dollars) needed in new plants and in waste discharge control facilities; and (8) the marginal value of limited capital for investment.

Detailed nonlinear mathematical models of steam electric power generation and water use and waste treatment systems were used to calibrate the important components of the linear power model developed in this study. Detailed engineering descriptions of stack gas scrubbers, mechanical and electrostatic precipitators and coal gasification systems were used where nonlinear models were not available. An interdisciplinary team of economists and engineers was used to develop a nonlinear steam electric power model and a water use and waste treatment model at the University of Houston. In-depth industry reviews of the nonlinear steam-electric power generation model (EBASCO) and the water use and waste treatment model (Fluor) developed in this project were obtained. The M.W. Kellogg study for the Environmental Protection Agency was the primary engineering basis for the stack gas scrubbing modules (Shore et al., 1974.)[2] A study by United Aircraft was the primary engineering basis for the coal gasification combined-cycle (CGCC) module (Robson et al., 1970).[3] Additional in-depth industry reviews of the linear CGCC model (Fluor) were obtained. Review suggestions from the Envirosphere Company, Houston Lighting and Power, and Consolidated Edison were considered in writing the final report.

A new feature of the linear electric power model is that it synthesizes the important technical data into a comprehensive economic basis for evaluating the effects of possible policy decisions on the use of resources, the availability of capital, the discharge of wastes, the costs of production, and the economic value of resources before policy decisions are made. This basis can be used to evaluate the response of cost-conscious managers to changes in resource availabilities, waste discharge standards, capital limitations, output requirements, resource prices, and effluent taxes.

Electric utilities operate as regulated industries; thus, cost-minimization as an objective may not correspond perfectly with actual management behavior of a utility. However, cost minimization does give the most efficient use of resources and is a justifiable societal norm for the nation. The model developed in this study may be tailored in further work to evaluate the possible inefficiencies induced by utility regulation. Naturally, this tailoring would involve explicit consideration of "peaking" power units and regional factors.

Results of Analyses

Environmental Analysis

Analyses were made which imposed increasingly severe waste discharge restrictions on the linear electric power model. Electric power generation requirements and fuel availabilities were set at 1985 projected levels. From a base of minimal discharge limitations, restrictions were placed individually on water discharges and on air emissions; restrictions were then placed on both air and water discharges together to determine the effects of these restrictions on production costs, production mix, resource use, waste discharges, and investment requirements.

Production costs increased by 50% from 1.27¢/kWh in the base case to 1.9¢/kWh in the most restrictive case. An intermediate case, roughly equivalent to EPA's 1977 Best Practicable Technology standards (BPT), resulted in production costs of 1.57¢/kWh. As environmental restrictions increase, energy use increases by 5%, water withdrawals decrease by 98%, water pollutants decrease by 100%, and air pollutants decrease by 87%. The production mix changes from high investments in new high-sulfur coal-burning plants in the base case to heavy investments in coal gasification combined-cycle plants (using high-sulfur coal) in the most restrictive case. This use of CGCC indicates a shift from desulfurizing stack gas emissions to the removal of sulfur from coal prior to combustion. Cumulative capital investments required to accomplish this shift in technology by 1985 total nearly $68 billion.

Electricity Supply Analysis

The economic supply of electric power and the technological configuration required to produce it are sensitive to fuel and capital availabilities and to waste discharge restrictions. Combinations of low electricity requirements, high availability of clean fuels for power generation, and 1974 waste discharge standards result in (1) continued use of natural gas and high-sulfur coal to fire steam boilers, (2) low marginal and average costs of producing electricity, and (3) positive marginal costs of waste discharge control only for particulates. Simultaneously increasing the requirements for electricity, increasing the restrictiveness of the effluent standards, and decreasing the availability of the clean fuels results in (1) heavy investments in gasification of coal and combined-cycle power generation, (2) high marginal and average costs of producing electricity, (3) high marginal values of clean fuels, (4) high marginal costs of controlling air and water pollutants, and (5) high requirements for investment capital. The following ranges of production costs, fuel values, and waste discharge control costs were found: (1) average costs of electricity, 1.3¢ to 2.51¢/kWh; (2) marginal costs of producing electricity, 1.03¢ to 2.83¢/kWh; (3) marginal value of natural gas, zero to $3.19/MMBtu; and (4) marginal costs of sulfur dioxide control, zero to $0.16/lb.

Factor Demands for Water Withdrawals and Low-Sulfur Coal

Water. The mix of generation processes was found to be sensitive to the requirements for electricity, the availability of clean fuels, the availability of capital, and the restrictiveness of the waste discharge standards; however, the mix of generation processes was found to be invariant to the price of water withdrawals in the range of prices from zero to $4.00/Mgal. Indications are that the production and the water use systems are economically separable with respect to water policy evaluations.

Increasing the price of water withdrawals resulted in significant changes in the water use system. Wet cooling towers were used at relatively low withdrawal prices of around 1.5¢/Mgal; water withdrawal decreased 97% to slightly less than 1 gal./kWh. Water reuse was initiated at withdrawal prices of approximately 22¢/Mgal; complete water reuse was accomplished at water withdrawal prices ranging from $3.21 to $6.79/Mgal.

A comparison between the absolute standards on discharge of pollutants and use of water withdrawal price increases reveals considerable divergence in these two approaches to water reuse. Increasing the water withdrawal price is not the most cost-effective way of achieving water reuse, although it is a means of controlling withdrawals. The increase in water withdrawal price tends to limit not only water discharge but also in-plant water consumption. This means that a high water withdrawal price results in a much more stringent overall water policy than does the use of an absolute standard of zero discharge. For the purpose of achieving minimum discharge of water pollutants, the water withdrawal price mechanism results in larger capital requirements and higher operating costs than does the application of absolute discharge standards. For a base case electricity demand of 1.5 trillion kWh in 1985, comparable results in minimizing water pollutant discharges required operating costs of 1.36¢/kWh and investment capital of $6 billion due to the application of absolute standards, while the water withdrawal price mechanism required operating costs of 1.49¢/kWh and $11.4 billion capital. The price of water withdrawal necessary to cause minimum water pollution was $3.21/Mgal.

Low-Sulfur Coal. The economic demand for low-sulfur coal in electric power generation is sensitive to the availability of petroleum fuels, the price of low-sulfur coal, electricity requirements, and the restrictiveness of sulfur dioxide standards. With electricity requirements of 2.5 trillion kWh, relatively abundant availabilities of petroleum fuels, and implementation of legislated effluent standards, use of low-sulfur coal in the model was 11 quadrillion Btu's at a price of 61¢/MMBtu ($10.37/ton) and zero at all higher prices of low sulfur coal. At 61¢/MMBtu, decreasing the availability of petroleum fuels by 82% increased the use of low sulfur coal by 64%; increasing the restrictiveness of the waste discharge standards to require the use of stack gas scrubbers and zero discharge of pollutants to the water increased the use of low-sulfur coal another 17%; and increasing the level of electricity requirements from 2.5 to 3.3 trillion kWh increased the use of low-sulfur coal still another 31%.

With the availability of petroleum fuels at a minimum level needed by the model, the use of low-sulfur coal was relatively invariant to price in the range of 61¢ to $1.50/MMBtu. However, the use of low-sulfur coal was extremely sensitive to prices between $1.50 and $2.00/MMBtu. Use of low-sulfur coal in that price range decreased by as much as 75% in the case of the most restrictive waste discharge standards combined with the highest level of electricity requirements.

Capital Demand. Combinations of strict effluent standards and high production of electricity from nuclear power resulted in a maximum capital investment of $192 billion for the electric power industry from 1975 through 1985. Combinations of lax environmental standards and low production of electricity from nuclear power resulted in a minimum capital investment of $165.3 billion for the same period. These results of the model compare with a maximum forecast availability of $261 billion for the electric power industry and with a minimum forecast availability of $163 billion. The model shows that capital availability, if the minimum forecast materializes, could inhibit the ability of the electric power industry to meet electricity demand.

References

1. Federal Energy Administration, *Project Independence Report: Project Independence,* U.S. Government Printing Office, Washington, D.C., 4118-00029, Nov. 1974.
2. D. Shore, J.J. O'Donnell, and F.K. Chan, *Evaluation of R&D Alternatives for SO_x Air Pollution Control Processes,* prepared for Office of Research & Development, by the M.W. Kellogg Co., U.S. Environmental Protection Agency, Washington, D.C., Sept. 1974.
3. F.L. Robson, A.J. Giramonti, G.P. Lewis and G. Gruber, "Technological and Economic Feasibility of Advanced Power Cycles and Methods of Producing Nonpolluting Fuel for Utility Power Stations," United Aircraft Research Laboratories, East Hartford, Conn., 06108, Dec. 1970, Final Report, Contract: CPA 22069-114. UARL Report J-970855-15. Prepared for National Air Pollution Control Administration, U.S. Dept. of HEW.

1. The Linear Electric Power Model

L. Ted Moore ▪ James A. Calloway
Russell G. Thompson

A linear economic policy model of electric power generation was developed to assist policymakers in answering questions concerning the effects of stringent air emission and water discharge restrictions as well as limited capital and fuel availability on (1) the configuration of the power production complex, including generation mix and waste treatment processes used; (2) generation costs; (3) fuel mix; and (4) the value of clean fuel. The model can reflect the least-cost technological generation mix, fuel mix, and waste treatment configuration for any specified base year for the nation, a state, or a region.

Model Description

A conventional steam electric power plant is composed of several interrelated processes. These processes include power generation, heat removal, primary water treatment, wastewater cleanup, and stack gas cleanup (*Power,* 1967).[1] Figure 1.1 illustrates a typical configuration for fossil-fueled steam-electric power generation. A mathematical description and complete data listing for the model is included in Appendix A.

To produce electric power, fuel is used to make steam which drives a steam turbine/generator. The exhaust steam from the turbines is condensed, returned to the boiler, and reconverted to steam. Water supplied to the steam boiler must be very clean; that is, suspended solids, dissolved solids, and silica must be removed. However, the water treatment processes used to prepare boiler-quality water produce waste streams which contribute to the plant's water pollution problems. Facilities must be provided for disposal of these waste streams if stringent wastewater controls are applied.

The economic model of electric power generation includes a variety of alternatives for generating electric power from fossil fuels. All modeled units presently exist or could be operational by 1985. The alternatives include both old and new steam electric and combined cycle plants which collectively use a variety of fuels. New plants are assumed to have increased efficiencies over old plants. The fuel mix includes low- (0.75%), medium- (2.4%), and high-sulfur coal (6%); low- (0.5%) and high-sulfur oil (2%); and natural gas. Coal

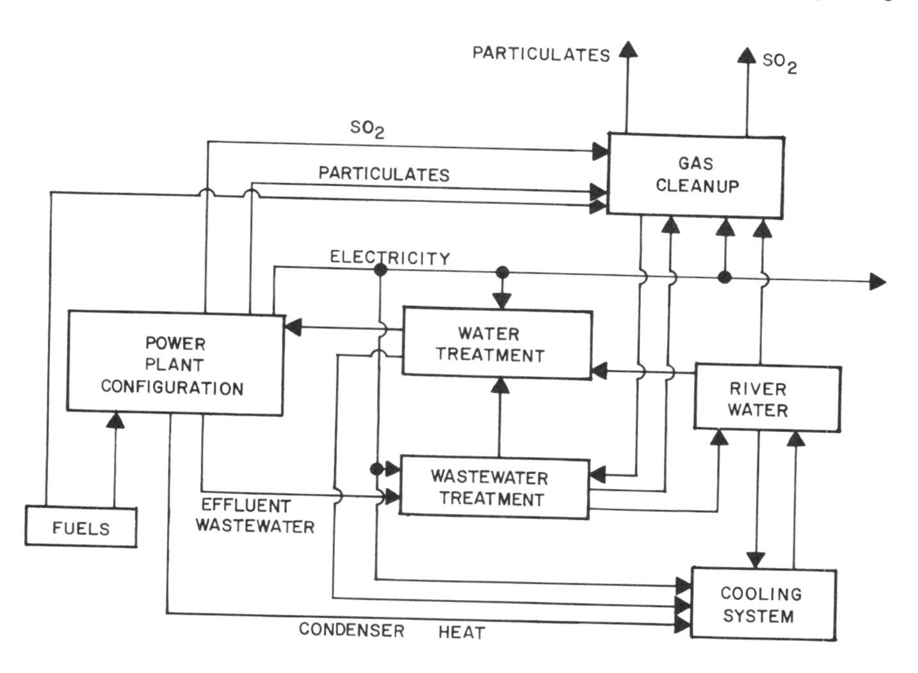

Figure 1.1. Linear electric power model flow diagram.

gasification of low-, medium-, and high-sulfur coal is also included to provide an alternate source of fuel in new integrated coal gasification-combined cycle plants; this technology is discussed in Chapter 2. In addition, it is assumed that old non-coal-burning plants can switch to coal to fire boilers.

Waste treatment alternatives are provided in the model for control of air emissions (particulates and sulfur dioxide) and water discharges (dissolved solids, suspended solids, and heat); waste discharges resulting from stack gas cleanup are also controlled. These air emission and wastewater control units, their operating characteristics, and specific use within the model are discussed in Chapter 3.

Under a given set of production requirements, fuel availabilities, and effluent discharge constraints the model may choose among any of the production and waste treatment technologies to meet the demand for electric power. Table 1.1 lists the alternative production processes available to the model and the corresponding heat rate assumptions applicable to the various types of plants (National Coal Association, 1974).[2] The heat rate for a plant is the amount of fuel energy input required to produce 1 kWh of electric power. The least efficient plants of the group are the existing oil-burning plants (31%); the most efficient are the new combined-cycle plants (48%). However, low heat

Table 1.1. Heat Rate Assumptions for Old and New Plants

Type of Plant	Heat Rate (Btu/kWh) Old	New
Coal-burning	10,250	8600
Oil-burning	10,880	8500
Gas-burning	10,840	8500
Combined-cycle	8000	7000
Coal gasification combined-cycle	—	8900

rates by gas- and oil-fueled plants are somewhat deceptive in that these are overall average heat rates which include fuel use in peaking units which have very high heat rates.

Table 1.2 lists the resource inputs (indicated by a minus) and product outputs for models of an old coal-burning plant and a new coal-burning plant. Entries in the table are based on production of 1 kWh of electrical energy. Plant model costs are developed by treating all generation as base load generation. Current operations by utilities consist of meeting three distinct types of loads—base, mid-range, and peak (Humphries and Guild, 1971).[3] The hardware assigned to each category of generation loading is defined as follows:

Base load—operating 5000 hours or more per year
Mid-Range—operating 1000 to 5000 hours per year
Peaking—operating 1000 hours or less per year

Table 1.2. Model Coefficients for New and Old Low-Sulfur Coal-Burning Electric Power Plants

	Old Plant	New Plant
Variable costs ($)	0.005	0.0063
Electric power (kWh)	1.0	1.0
Low-sulfur coal (Btu)	−10,250	−8600
Boiler water (gal.)	−0.1	−.07
SO_2 produced (lbs.)	.0171	.0129
Stack heat produced (Btu)	1200	1000
NO_x produced (lbs.)	0.0293	0.0246
Particulates produced (lbs.)	.1532	.1155
Waste heat produced (Btu)	5400	4100
New construction cost (¢)	—	3.83

A distinction in terminology must be made here. Load duration curves for electric utilities generation indicate high generation requirements called "peak demand" in the summer and winter. A large part of this demand may be met primarily by base load and mid-range units with peaking units contributing only a small percentage of peak demand; so the meeting of peak load requirements does not necessarily imply the use of units defined as peaking units. In fact, U.S. utilities had a reserve margin of capacity equal to 25.1% of peak load demand during the summer peak of 1973 (FPC, 1975).[4] Separate accounting for mid-range and peaking capacity has not yet been included in this model. Instead, these various generation types have been accounted for by using the overall average heat rate for units using the same fuels and treating all such units as base load units. Using this approach, old units were treated as base load units operating at an average plant factor of 50% (Shore et al., 1974).[5] The plant factor is defined as (net generation in kWh x 100) divided by (plant capacity in kW x 8760).

A variety of controls can be placed on the operation of the plants. For example, old low-sulfur coal-burning plants can be constrained to produce, within a range of values, a predetermined proportion of the total requirements for electric power. Similar constraints can be placed on all old production capacity. For the analyses discussed in this report the various types of old plants were constrained to operate at no higher level than they did in 1974 except for the provision that old gas- and oil-burning plants could switch in large part to the use of coal.

The model allows a large variety of permutations of modeled hardware to be used to meet generation requirements. In general, use of equipment which would be required solely for the purpose of meeting very restrictive air or water standards is optional. Devices which have this optional feature in the model are particulate precipitators, sulfur abatement equipment, evaporators, wet cooling towers, and dry cooling towers. By allowing the model to use this type of hardware only as policy dictates, the capital costs of adding this equipment to the generating units can be obtained from the "new construction" totals in the model output. By restricting the amount of capital available, the model can be made to select hardware combinations which satisfy the dual criteria of least operating cost and limited capital.

Overall, the model is able to select generating equipment on a least-cost basis as a function of fuel availabilities, fuel costs, pollution policy constraints, and available capital. A complete data listing of the model is provided in Appendix A.

Analyses may be conducted under a variety of operational and environmental scenarios, but for a typical base case, the model has been set up to produce one year's electric power requirements for the year 1985. Under this assumption, investment in new production equipment and/or waste treatment is influenced by: (1) limitations on the amount by which generation by old production equipment may be reduced by the base year; (2) limits on the availability of certain types of fuels, especially low-sulfur fuel; and (3) limits on the availability of investment capital.

Capital Investment and Operating Costs

Capital is treated as a resource necessary for the operation of new units in the model. Availability of this resource can be restricted in the same way as any other resource. When the supply of capital available to the model is set at a low value (relative to the amount which the model ordinarily uses), process substitutions occur. Other effects of a scarcity of capital include changes in the fuel mix which the model requires and increases in operating costs. Analyses of the model's capital requirements patterns are discussed fully in Chapters 4 and 9.

To allow the evaluation of capital requirements by use of the model, capital coefficients are included in the tableau of coefficients representing the model. In each column representing a new facility, a capital cost coefficient is included. However, old (existing) facilities do not include capital requirements, so the columns corresponding to these old units do not include a capital requirements coefficient. For example, since it is assumed that all old coal plants in the model are equipped with particulate control units, no capital costs are assigned to these units; but since stack gas scrubbers are assumed *not* to be in place, all scrubbers in the model have a capital cost assigned to them. In fact, scrubbers operating at old plants have a higher capital cost assignment than those at new plants because of the additional construction costs required to modify old plants.

A yearly fixed capital charge of 14% of plant capital costs is assumed to be a part of the annual operating costs of each unit of equipment in the model. Other operating costs were estimated from historical data (FPC, 1975).[4]

Both annual operating costs of a unit and its capital costs (where applicable) are prorated over the yearly output of the unit's operation. For example, consider the model of a new low-sulfur coal-burning plant with a production capacity of 1000 megawatts. Construction costs for this unit (in 1973 dollars) are placed at $230/kW (United Engineers and Constructors, 1972).[6] Assuming an annual operating load of 6000 hrs/yr, the prorated capital costs amount to 3.83¢/kWh of generation. A similar calculation of operating costs produces a figure of 6.3 mills/kWh.

The total capital required by a new unit is added to the capital requirements of all other new units. The capital requirements produced by the model solution represent total capital necessary from 1974 through the specified year of analysis, which in this study is 1985. If no new units were required to the year being investigated, the capital total would be zero.

Except for the analyses in Chapter 9, all analyses in this report are directed toward investigating the generating mix of fossil-fueled plants and their auxiliary units, exclusive of transmission and distribution hardware. The capital assumptions for new process equipment are listed in Table 1.3.

Table 1.3. Capital Cost Assumptions for New Production Equipment

Process	Capital Costs ($/kW)
Coal-burning plants	230
Oil-burning plants	228
Gas-burning plants	200
Combined-cycle plants	180
Coal gasification combined-cycle	254
Cooling tower	7
Scrubber (new plant)	40
Scrubber (retrofitted to existing plant)	70
Particulate treatment unit	5.50

References

1. Editors of *Power, Power Generation Systems,* McGraw-Hill, Inc., New York, N.Y., 1967.
2. National Coal Association, *Steam-Electric Plant Factors,* Washington, D.C., 1974, p. 102.
3. James J. Humphries, Jr. and Donald H. Guild, "Technical and Economic Features of Stone and Webster's FAST Combined-Cycle Power Plants," *American Power Conference 33,* 1971. pp. 492-502.
4. Federal Power Commission, *1974 Annual Report,* Washington, D.C., pp. 77-78.
5. D. Shore, J.J. O'Donnell, and F.K. Chan, "Evaluation of R&D Alternatives for SO_x Air Pollution Control Processes," prepared for Office of Research and Development, by the M.W. Kellogg Co., U.S. Environmental Protection Agency, Washington, D.C., 1974, p. 13.
6. United Engineers and Constructors, "100-MWE Central Station Power Plants Investment Cost Study," Volumes 1 through 5, prepared for USAEC, Division of Reactor Development and Technology, Contract No. AT(30-1)-3032, June 1972, WASH-1230.

2. Coal Gasification — Combined-Cycle

L. Ted Moore ▪ James A. Calloway
Russell G. Thompson

Presently, combined-cycle electric power generation is being accepted as an efficient, commercially proven technology. However, the unavailability of clean fuel (natural gas) to drive the gas turbine generators is a major problem. One promising method for providing a reliable fuel is coal gasification. Coal gasification combined-cycle (CGCC) power plants provide a method by which an abundant fuel (coal) can be used to generate electric power with minimum detriment to the environment (Robson et al., 1970).[1] Further, CGCC power plants can be operational by 1985 or sooner.

An explanation of the CGCC system requires discussion of the similarities and differences between conventional steam electric plants and coal gasification combined-cycle (CGCC) power plants.

A conventional fossil-fuel fired steam electric power plant is illustrated in Figure 2.1. The facility converts fuel to steam in a high-pressure boiler. The steam drives a steam turbine which, in turn, drives a generator to produce electricity. Condensate from the turbine is recirculated to the boiler. Many kinds of fuels may be used to fire the boiler, but if environmental restrictions regulate discharges to the air, a clean fuel such as natural gas, low-sulfur oil,

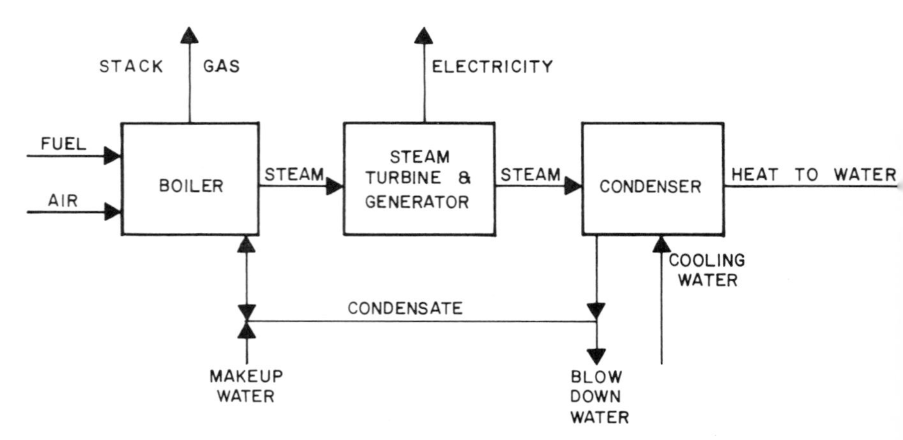

Figure 2.1 Power plant diagram for a conventional steam electric plant.

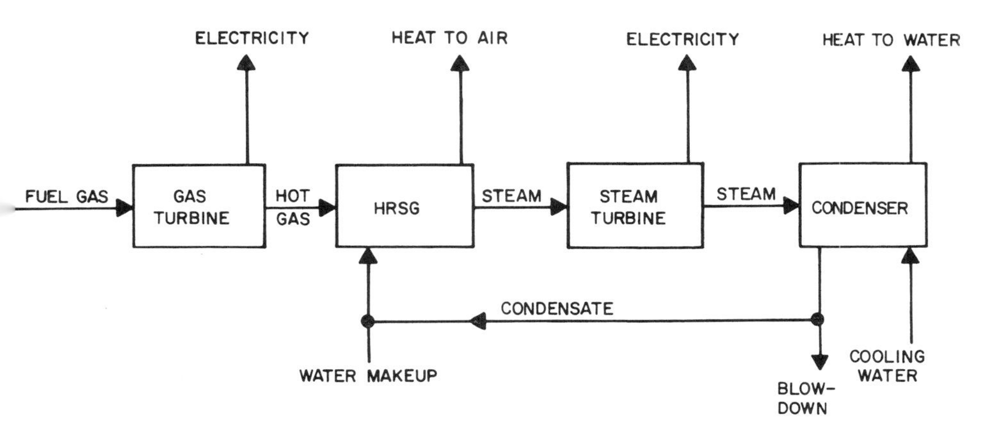

Figure 2.2. Power plant diagram for a combined-cycle electric plant.

or low-sulfur coal is preferred. To comply with any imposed air restrictions, the plant has the option of (1) using a clean fuel, (2) cleaning the fuel before use, or (3) cleaning the stack gas after combustion.

The combined-cycle plant illustrated in Figure 2.2 burns a clean fuel (usually natural gas) in a gas turbine generator to produce electricity. Exhaust heat from the turbine is recovered by a heat recovery steam generator (HRSG). Steam is produced in this heat recovery system and used to drive a steam turbine generator to produce additional electricity. To avoid corrosion and erosion of the turbine blades, the fuel used in a combined-cycle system must be clean, that is, sulfur-free gas or oil. Thus, the plant has only the option of using a clean fuel or cleaning the fuel prior to use. A complete CGCC system integrates the combined-cycle power plant with coal gasification, gas cleanup, and desulfurization to generate electricity using a clean fuel produced from coal. Development of the CGCC submodel was based upon a 1000-MW capacity plant.

Efficiency

The primary measure of the efficiency of an electric power plant's operation is its "heat rate"; that is, the number of Btu's of fuel heating value required to produce 1 kWh of electricity. This heat rate can be converted to an efficiency factor by taking the ratio of the heat equivalent value of a kilowatt hour (3413 Btu/kWh) to the heat rate of the plant. For example, a heat rate of 10,252 Btu/kWh translates into an operating efficiency of 33.3%, the U.S. average efficiency for coal-fired plants. Some combined-cycle plants have demonstrated operating efficiencies up to 42%, while the highest current steam electric efficiency has been about 39% (Wood, 1972).[2]

Gas turbine efficiency is primarily a function of the inlet gas temperature; technological improvements have resulted in a steady rise in allowable inlet temperatures such that gas turbine efficiencies as high as 27% are projected for the near future (General Electric, 1973).[3] The overall efficiency of the combined-cycle plant is a function of the efficiency of the gas turbine unit, the electric power output of the gas turbine unit and the steam electric unit. That is,

$$\eta_{CC} = \eta_{GT} \left(1 + \frac{E_{ST}}{E_{GT}}\right) \qquad \text{(Moore, 1975)}^{4} \qquad (1)$$

where η_{CC} = conversion efficiency of the combined cycle plant
η_{GT} = conversion efficiency of the gas turbine
E_{ST} = electrical output of the steam electric unit and
E_{GT} = electrical output of the gas turbine unit

The linear model assumes a steam turbine/gas turbine power output ratio of 0.74 and a gas turbine efficiency of 27%; the overall efficiency of the combined-cycle system is calculated to be 47%. Assuming the coal gasification unit has a thermal efficiency of 82%, the CGCC system has an overall efficiency exceeding 38%, corresponding to a heat rate of 8900 Btu/kWh. This efficiency compares to that of a conventional steam electric plant with stack gas clean-up, expected to be less than 35% (Foster-Pegg, 1966).[5]

Coal Gasifier

Coal is gasified by heating it under a specific set of temperature and pressure conditions and in the presence of steam and air or oxygen. Most of the heat is supplied by the exothermic reactions of the coal with oxygen or air. Most of the sulfur and impurities contained in the crude gas must be removed before the gas is fed to the gas turbine.

The best proven gasifier process presently available is the Lurgi gravitating bed gasifier. In this process, coal is crushed and fed into a lock hopper above the gasifier after removal of fines. The coal is then heated to a temperature just below the ash fusion point (normally 2000-2800°F). Steam and either air or oxygen at approximately 22 atmospheres of pressure are introduced at the bottom of the gasifier. Use of air produces a nitrogen-containing gas with a heating value of 173 Btu/ft.^{3} (compared to 1050 Btu/ft.^{3} for natural gas); using oxygen, the heating value is 445 Btu/ft.^{3}. Either of these gases can provide sufficient inlet gas temperature for the gas turbines in the combined-cycle unit (see Figure 2.3), and, since the airblown process is cheaper, its use has been assumed in the CGCC model.

The complete gasification system includes part of the gas cleanup section. This cleanup is a quench-water scrubbing step which removes tar, dirt, and lkalies from the crude gas (*Combustion,* February, 1970).[6] Tar recovered in he quench step is used to briquette the fines for use in the gasifier; dry ash is emoved from the bottom of the gasifier for disposal. The waste heat boiler emoves sensible heat from the crude gas to produce part of the gasifier rocess steam. The major portion of the steam is produced in the main rocess boiler fired by natural gas, low-sulfur oil, or a portion of the esulfurized gas from the gasifier output.

The linear programming CGCC model is based upon a 1000-MW CGCC lant and assumes the use of 23 such gasification systems which produce a as of approximately 158 Btu/ft.[3] heat content before desulfurization and 173 tu/ft.[3] after desulfurization. The heating value of the output gas is approx- nately 82% of that of the inlet coal (Robson et al., 1970).[1]

esulfurization

The primary constituents of the gasifier gas are methane (CH_4), water H_2O), carbon dioxide (CO_2), carbon monoxide (CO), hydrogen (H_2), itrogen (N_2), and hydrogen sulfide (H_2S). Of these, only H_2S causes a roblem (combustion of H_2S produces SO_2, which corrodes turbine parts) 1at necessitates its removal.

Two processes are currently favored for sulfur removal: the amine process nd the Benfield hot potassium carbonate process. The hot potassium car- onate process is generally more reliable, requires less steam and solution 1akeup, and is less expensive to operate (Benson and Field, 1960).[7] The near programming model assumes the use of up to six hot potassium car- onate process units in a 1000-MW plant. The H_2S is absorbed in an ab- orber and then steam-stripped in a stripper and sent to a Claus sulfur ecovery system; the regenerated potassium carbonate is recycled to the ab- orbing portion of the system. Elemental sulfur is recovered as a salable roduct from the Claus plant. Residual gas from the Claus unit is either in- nerated (which produces some effluent SO_2) or cooled, compressed, and ecycled to the Benfield cleanup step.

Two types of sulfur recovery units were included in the model: (1) a Claus lant with incineration of the tail gas and (2) a Claus plant with recycle of the il gas. The latter process is assumed to completely eliminate SO_2 emissions om the desulfurization plant (see Figure 2.3).

as Turbines

The LP model assumes use of three gas turbine electrical generating units, ach with a new capacity after in-plant consumption of 193 MW. Operating ata are based upon a CGCC plant with a net output of 1000 MW. The air ompressor, gas turbine, and electric generator are all attached to the same

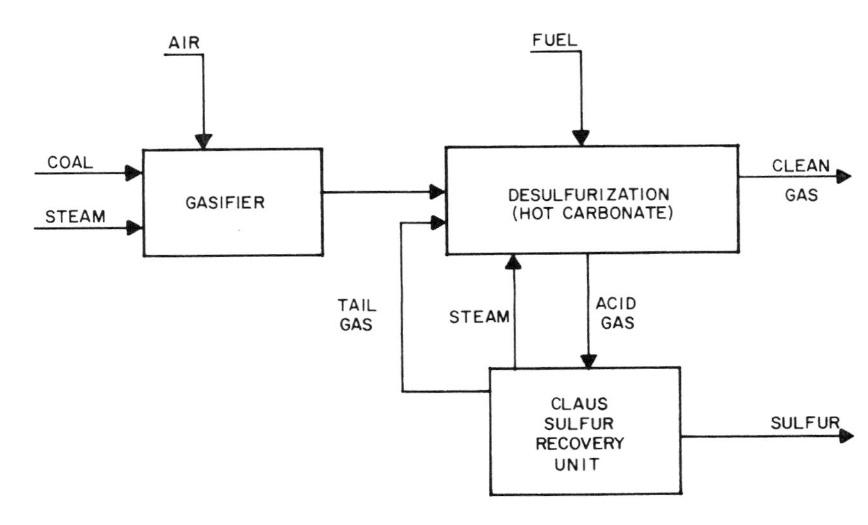

Figure 2.3. Gasifier and desulfurization processes.

shaft. Gas enters the combustion chamber with compressed air. The gas-to air ratio is determined by the desired temperatures of the burned gases, in thi case 2200-2300°F. The hot exhaust gases from the turbine (about 1300°F are routed to the heat-recovery boiler section of the steam-turbine systen described below. Assumed thermal efficiency of the modeled gas turbine i 27%.

Steam Turbines

The steam electric portion of the combined-cycle plant includes a hea recovery boiler (HRSG), steam turbine, electric generator, and a steam con denser. The HRSG includes a reheater, vaporizer, and an economizer an two superheaters. The economizer heats water to the boiler and the steam i formed in the vaporizer. The LP model assumes the use of two superheater to produce steam at 2400 psia and 1000°F, with a single reheat cycle t 1000°F. The net capacity of the steam turbine generator is 431 MW; th steam condenser is assumed to operate at 2 inches of mercury and 101.4°F

Cost Estimates

Capital costs are assumed to be $58 million for the gasification plant, $8 million for the gas turbines, $21 million for the desulfurization plant, and $9 million for the HRSG plant for a combined figure of $254/kW, assuming maximum gas cleanup. Capital costs for the desulfurization plant without ta gas recycle are $20 million. The yearly cost of capital was assumed to be 14% of the plant's nominal capital value. A cost of 5% of capital was assumed for maintenance and 6% of capital for labor cost. The CGCC plant factor was assumed to be 60%.

References

1. F.L. Robson, A.J. Giramonti, G.P. Lewis, and G. Gruber, "Technological and Economic Feasibility of Advanced Power Cycles and Methods of Producing Nonpolluting Fuel for Utility Power Stations," United Aircraft Research Laboratories, East Hartford, Conn., 06108, Dec. 1970, Final Report, Contract: CPA 22-69-114, UARL Report J-970855-13. Prepared for National Air Pollution Control Administration, U.S. Dept. of HEW.
2. B. Wood, "Combined Cycles: A General Review of Achievements," *Combustion*, Apr. 1972, pp. 12-22.
3. G.A. Ludwig, "General Electric Gas Turbine Design Philosophy," Gas Turbine Reference Library Paper GER-2483, General Electric Co., Schenectady, N.Y., 1973.
4. L. Ted Moore, *An Economic Model of a Coal Gasification-Combined Cycle Electric Power Plant,* Ph.D. Dissertation, Rice University, Houston, Texas, June 1975.
5. R.W. Foster-Pegg, "Gas Turbine Heat Recovery Boiler Thermodynamics, Economics and Evaluation," *Combustion*, Mar. 1971, pp. 8-18.
6. "Power Plant Integrated with Pressure Gasification of Coal (Lurgi)," *Combustion*, Feb. 1971, pp. 12-13.
7. H.E. Benson and J.H. Field, "New Data for Hot Carbonate Process," *Petroleum Refiner*, April 1960, pp. 127-132.

3. Water Use and Waste Treatment

L. Ted Moore ▪ James A. Calloway
Russell G. Thompson

The water use and waste treatment systems described in the linear electric power model consist of those processes necessary to assure minimum discharge of both waterborne and airborne pollutants. The primary water pollutants are dissolved solids, suspended solids, and heat. Both dissolved and suspended solids are produced through primary and secondary treatment of raw inlet water for process use. Heat discharges to water result from the use of once-through cooling. Airborne pollutants of particulates and SO_2 are residues from burning fuel.

The linear model allows treatment of waterborne wastes to achieve zero discharge of these wastes through clarification, demineralization, evaporation, and sludge handling for a wide variety of fuel use and technology combinations, including both old and new installations. Maximum possible removal of airborne wastes is achieved using precipitators and stack gas scrubbers. The latter, however, contribute to the overall waste treatment problem in that current air emissions control technology tends to convert air pollution problems to water pollution problems and solid waste problems.

Wastewater Discharges

Figure 3.1 depicts the water use and waste treatment system described by the linear model. The waste treatment model is derived from a more general water use and waste treatment model developed by F.D. Singleton (1975).[1] This model accounts separately for each major component of the cooling water and wastewater streams and the component removal efficiencies of each waste treatment unit in the model plant. Nonlinear water use and waste treatment models are used to estimate the costs of each waste treatment unit.

Water is assumed to be available from a nearby river. Raw river water is used to provide once-through cooling, ash removal, and makeup to the stack gas scrubber. River water is clarified using lime and alum to remove suspended solids and to produce makeup water for recycle cooling towers and wash water for utility boilers. Finally, clarified water is demineralized to remove dissolved solids. This process produces very pure water which is used in the utility boilers to make steam.

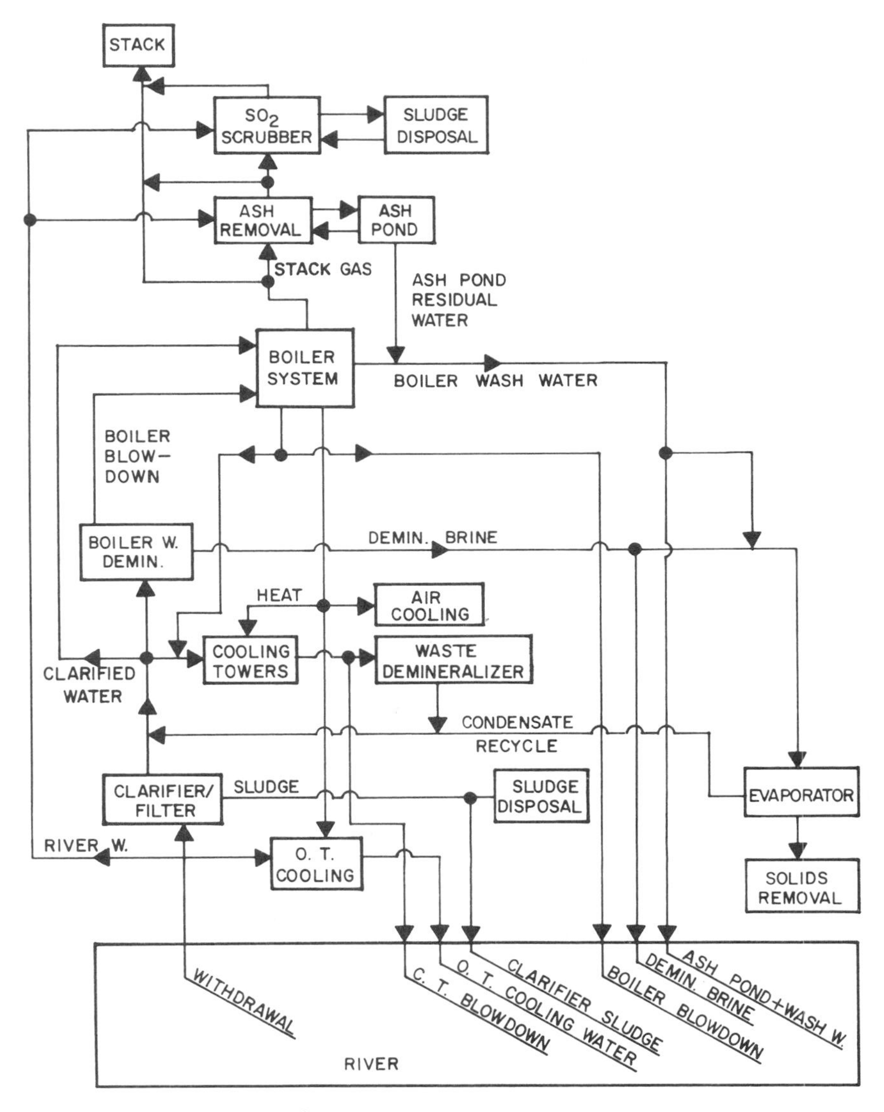

Figure 3.1. Water use and waste treatment system.

In the model, the inlet water is the major source of suspended and dissolved solids. This water is assumed to be of medium quality for the nation and contains suspended solids, dissolved solids, and silica. Suspended solids are removed in the clarifier. The model allows the sludge produced in the clarifier to be discharged to the river or dewatered and sent to landfill.

High-pressure steam boilers used in electric power generation require very pure inlet water, that is, water from which ionic impurities such as sodium, calcium, magnesium, sulfate, chloride, and silica have been removed. These dissolved solids are removed by an ion-exchange demineralizer utilizing beds of anion and cation exchange resins.

The resins must be regenerated periodically; either sulfuric or hydrochloric acid regenerates the cation bed and caustic soda regenerates the anion bed. Unfortunately, the regeneration process adds approximately 1 pound of solids to the demineralizer waste stream for each pound of solids removed from the inlet stream. The resulting brine stream may be discharged to the river or evaporated to dryness. Condensate from the evaporator is recycled to the process; soluble salts from the evaporator are disposed of in a specially prepared landfill (General Technologies Corp., 1973).[2]

Fresh demineralized water, equal to about 1% of the steam generation flow rate, is added to the boiler system as makeup. An equal volume of water is withdrawn from the system, as boiler blowdown, to remove trace impurities. This blowdown is either discharged to the river or recycled as cooling tower makeup. Wash water which is used periodically to clean the boilers is either discharged or evaporated.

The model provides for three alternative cooling methods: once-through cooling, wet recycle cooling, and air cooling. Raw river water is used for once-through cooling; clarified water is used to make up the evaporation losses from the cooling tower. Use of cooling towers virtually eliminates discharges of heat to the water source but produces an effluent, called blowdown, which contains the dissolved solids and silica originally contained in the input water. The model provides that this blowdown may be discharged to the river or demineralized and reused, as described above.

In both the ash removal and stack gas scrubbing steps, sludges of suspended solids are discharged into settling and evaporation ponds. Suspended solids from the ash removal accumulate on the bottom of the ash pond and the water is recycled for reuse. Sludge from the scrubber will not settle out, and that sludge is dried in an evaporation pond.

Dissolved solids concentrations in the various water streams depend on local river water quality, type of fuel used, plant design, and plant operating methods. For modeling purposes, the stream concentrations shown in Table 3.1 have been assumed. In the model analyses discussed in Chapter 8, water quality was varied in the model over the likely range of U.S. water quality at steam electric plants. The results of these analyses are discussed in Chapter 8, but the changes in plant operating costs due to changes in water quality were found to be relatively small.

The water treatment system is considered an integral part of the overall power plant; therefore, the capital costs of the water treatment system are included in the total plant costs shown in Table 1.3.

Table 3.1. Dissolved Solids Concentrations Assumed for the Linear Model

River and clarified water	350 mg/liter
Cooling tower blowdown	1750 mg/liter
Boiler blowdown	50mg/liter
Ash Pond residual water	5000 mg/liter
Boiler wash water	5000 mg/liter
Demineralizer waste brine	8000 mg/liter if for discharge; 2.5% if for evaporation

Summarizing briefly, under the assumption of a nonrestrictive waste discharge policy, all wastewater streams are discharged to the natural watercourse. If, however, discharge restrictions such as 1983 BPT standards are imposed, (1) clarifier sludge is dewatered and sent to landfill, (2) cooling tower blowdown is demineralized and reused in the cooling tower, (3) demineralizer brine, ash pond residual water, and boiler wash water are evaporated in an evaporator/dryer system, (4) scrubber sludge is evaporated to dryness in a pond, and (5) dry solid residues are sent to landfill.

Air Emissions

A steam electric plant produces two primary types of air emissions: particulates and sulfur oxides. The most common of these oxides is sulfur dioxide (SO_2). Particulates are produced when the plant uses coal or fuel oil as boiler fuel. The emphasis on air pollutants in the model was placed solely upon SO_2 and particulate removal. SO_2 is released when the boiler fuel contains sulfur. There is no currently acceptable technology which will completely remove these constituents from the stack exhaust; however, a high percentage of particulate removal can be accomplished using mechanical and electrostatic precipitators.

Particulate Removal

In the linear model, control of particulate emissions from steam electric plants is accomplished in two stages (Walker, 1968).[3] The first stage is a cyclone precipitator which circulates the power plant exhaust gases to remove large particles of ash. The cyclone swirls the gases in a spiraling motion, causing the larger particles to strike the extruding walls and drop into a particle collector. The precipitate is removed from the system in a wash water and sluiced to a settling pond.

An electronic precipitator is used in the second stage to remove small particles. A large number of charged electrodes are used in the electrostatic precipitator to attract the fine particles. Periodically, the attached fines are knocked loose from the electrodes to fall into the particulate collector. A wash water flow removes the fines from the system and transports them to a settling pond. Surface water from the settling pond is recycled to the precipitator, and the dried sludge is used for landfill. The particulate removal efficiency assumed for this two-stage process is 97%.

Sulfur Removal

The simplest method for eliminating sulfur oxide emissions is to burn a low-sulfur fuel. In fact, this is the method most electric power plants have used in the past. However, because of the growing scarcity of low-sulfur fuels, alternative methods are being developed. Coal is the most abundant alternative fuel; therefore, the SO_2 control alternatives include use of sulfur-containing coal in conjunction with some form of sulfur removal technology. The available removal technologies include (1) sulfur removal prior to combustion, (2) coal gasification with gas purification, and (3) stack gas scrubbing.

The pre-combustion procedure is capital intensive and has a high operating cost. Further development work appears necessary before large scale commercial application can be made. Coal gasification holds a great deal of promise and is discussed in Chapter 2. Some form of stack gas scrubbing technology will likely meet the short-term desulfurization needs of the electric power industry (EPA, 1974).[4]

A number of processes have been proposed for the removal of sulfur oxides from stack gases. These processes are classified as "regenerative" or "throw-away" according to whether the residue produced by the process can be sold or must be discarded. In either case, flue gas containing sulfur oxide is contacted with an aqueous alkaline solution which reacts with the sulfur oxide. In the regenerative types, stack gas SO_2 is first concentrated, then converted either to elemental sulfur or sulfuric acid. In the throw-away type, insoluble sulfates and sulfites of calcium are formed which must be disposed of as a wet sludge.

The Environmental Protection Agency has generally recommended use of any of three processes—lime scrubbing, limestone scrubbing, or the Wellman-Lord process—for sulfur oxide removal. The first two of these processes are throw-away types while the latter is a regenerative type. However, users of these processes have complained of operational problems, reliability problems, and difficulties in disposing of residual sludge.

The wet limestone process, chosen for the linear model, uses direct contact between the stack gas flow and a limestone slurry to remove SO_2 from the stack gases. The hot flue gas is first cooled in a Venturi scrubber. Next, the flue gas SO_2 is removed in a turbulent contact absorber (TCA) with a removal efficiency of approximately 90%. The cleaned flue gas is then reheated to dry it and restore its buoyancy. Naturally, the reheating fuel must be low in sulfur, indicating the use of either natural gas or distillate (Shore et al., 1974).[5] Natural gas is assumed to be the reheating fuel in the linear model.

Cost Assumptions

Capital cost estimates for particulate removal units from the literature vary from as low as $1.20 per kW in 1970 (Robson et al., 1970)[6] to $22 per kW in 1972 (Federal Power Commission, 1975).[7] The average capital cost for particulate removal units was $4.22 in 1972 (Federal Power Commission, 1975).[7]

Investment and operating costs assumed for the particulate removal portion of the model are $5.50/kW and 2 mills/kWh, respectively.

Stack gas scrubbing costs are a function of the sulfur content of the coal, the type of process used, size of the plant boiler, and whether the unit is installed on a new or an existing plant. Investment costs for stack gas scrubbing can range from $40-$125/kW depending on whether the plant is new or existing. Operating costs range from 1 to 3 mills/kWh. If the scrubber is a throw-away type, sludge disposal can cost up to $17/ton. Operating costs assumed for the linear model were 3 mills/kWh. Investment costs were assumed to be $40/kW for scrubbers for new plants and $70/kW for scrubbers retrofitted to old plants. Sludge disposal cost was assumed to be $10/ton. Increased land acquisition costs resulting from an increasingly restrictive waste discharge policy are not explicitly delineated in the model.

These costs were incorporated into the linear model by assuming full load operation of 4380 hrs. for old plants and 6000 hrs. for new conventional plants and prorating the costs over a year. All cost estimates in this monograph are expressed in 1973 dollars.

References

1. F.D. Singleton Jr., James A. Calloway, and Russell G. Thompson, "An Economic Model of Water Use and Waste Treatment," *Proceedings of The Second National Conference on Water Reuse,* Chicago, May 4-8, 1975.
2. General Technologies Corporation, "Development Document for Effluent Limitations Guidelines and Standards of Performance, Inorganic Chemicals, Alkali and Chlorine Industries," for EPA under Contract No. 68-01-1513, pp. VIII-14-16; p. VIII-80; June 1973.
3. A.B. Walker, "Operating Principles of Air Pollution Control Equipment—Guidelines for Their Application," *Design and Operation for Air Pollution Control,* Metropolitan Engineers Council on Air Resources, New York, Oct. 24, 1968.

4. Report of the Hearing Panel—National Public Hearings on Power Plant Compliance with Sulfur Oxide Air Pollution Regulations, U.S. EPA, Washington, D.C., January 1974.
5. D. Shore, J.J. O'Donnell, and F.K. Chan, "Evaluation of R & D Alternatives for SO_x Air Pollution Control Processes," prepared for Office of Research and Development by The M.W. Kellogg Company, U.S. Environmental Protection Agency, Washington, D.C., September, 1974.
6. F.L. Robson, A.J. Giramonti, G.P. Lewis, and G. Gruber, "Technological and Economic Feasibility of Advanced Power Cycles and Methods of Producing Non-polluting Fuel for Utility Power Stations," United Aircraft Research Laboratories, East Hartford, Conn., 06108, Dec. 1970, Final Report, Contract: CPA 22-69-114, UARL Report J-970855-13. Prepared for National Air Pollution Control Administration, U.S. Dept. of HEW.
7. Federal Power Commission, Steam Electric Plant Air and Water Quality Control Data for the Year Ended December 31, 1972 Based on FPC Form No 67: Summary Report, U.S. Government Printing Office, Washington, D.C., March 1975.

4. Resource Allocations

H. Peyton Young ▪ James A. Calloway
Rodrigo J. Lievano ▪ Russell G. Thompson

The resource restrictions on electric power production are of three types: fuel availabilities, capital availabilities, and the availabilities of water and air for effluent discharge, i.e. *effluent restrictions*. The assumptions concerning each of these three types of constraints are discussed below.

Air Pollution Emission Restrictions

Abatement schedules for air pollutants from fossil-fuel burners have been drawn up by the appropriate federal and local authorities and are already in effect. These schedules are aimed at preserving the quality of our environment for protection of the public health. Concern for environmental quality, from both the public and private sectors, is clearly established in three important legislative acts—the Clean Air Act of 1963, the Air Quality Act of 1967, and the Clean Air Act Amendments of 1970. The latter gave birth to the Environmental Protection Agency, which has primary responsibility for determination of ambient air quality standards, approval of State Inspection Plans (SIP's) for existing sources, and the implementation of New Source Standards for new installations.

On April 30, 1971, the EPA issued national ambient air quality standards (both primary and secondary) which specified the maximum allowable average three-hour, daily, and annual concentrations of sulfur oxides, nitrogen oxides, and particulate matter which could be discharged. The primary standards, to be met by mid-1975, specify concentrations of pollutants which, if exceeded, would be detrimental to public health. The secondary standards define the levels of air quality judged necessary to protect the public from known or anticipated adverse effects of a pollutant. These secondary standards are to be implemented within a reasonable time as specified by EPA.

New source standards apply to fossil fuel-burning power plants constructed after August 1971. These standards explicitly restrict sulfur dioxide (SO_2) and nitrogen oxide (NO_x) emissions. Of these types of pollutants, only SO_2 is of great, current concern. New source emission standards for new boilers limit emissions to 1.2 lb./MMBtu of coal burned and 0.8 lb./MMBtu of oil burned as boiler fuels. NO_x is not controlled in the linear electric power model for either new or existing facilities.

The standards set forth in the State Inspection Plans will apply explicitly to existing sources. These standards are less stringent than those for new sources; for large installations, however, standards for existing sources tend to approach new source standards.

In several of the cases investigated with the linear electric power model, the environmental constraints imposed are defined as existing environmental conditions imposed on old plants and new source standards imposed on new plants. Existing conditions refer to those conditions in existence in 1974. Specifically, these conditions are interpreted as particulate control at all plants burning particulate-emitting fuels and a limitation upon SO_2 emissions representative of current emissions. Current emissions are estimated at 0.027 lb. of SO_2 per kWh.

Fuel Availabilities

The increasing demand for energy in all sectors and the need for environmental clean fuels to comply with the emission standards under the Clean Air Act have created a serious shortage of low-sulfur oil, low-sulfur coal, and natural gas. The Federal Energy Administration (1974)[1] estimates this deficit to be approximately 225 million tons of coal equivalent in 1975, 175 million tons in 1977, and 100 million tons in 1980. The reduction in the deficit from 1975 to 1980 assumes increased production of low-sulfur coal and increased utilization of flue gas desulfurization systems. The prospects for rapidly increasing the domestic supply of clean fuels depend on (1) whether existing emission standards continue to be enforced, (2) the continued development and acceptance of stack gas emission control technology, (3) the type of controls placed on strip-mining, (4) water availability in the new North Plains coal regions, and (5) other economic and institutional factors. The uncertainty concerning enforcement of existing emission standards on sulfur oxides has inhibited the development of new coal mines and the installation of flue gas desulfurization systems by electric utilities, since a widespread relaxation of emission restrictions would make the required investments unnecessary (Federal Energy Administration, 1974).[1] Even the removal of these uncertainties does not guarantee an increase in the supply of low-sulfur coal from strip mines. In the meantime, while these uncertainties are being resolved, utilities have been choosing to convert to oil (often imported) rather than relatively scarce and expensive low-sulfur coal or installing stack gas emission control systems. In the period 1970-1973, 19 million tons (per year) of coal-fired electric-power generation capacity was converted to oil (equivalent to approximately 208,000 bbls/day) (U.S. Congress, Senate Committee on Public Works, 1973).[2]

In our study, the effects of limited clean fuel availabilities on the costs and technical configuration of electric power generation were evaluated. This required estimation of the quantities of natural gas, high and low-sulfur oil, and high and low-sulfur coal available for electric power generation.

The base allocation of fossil fuels for electric power generation was obtained from a national study of supply, demand, and prices of fossil fuels (Thompson et al., 1975)[3] in which the electric power generation industry was a component of a large economic model of production and in which it bid for fuel along with the residential and commercial, industrial, and transportation energy-use sectors. Production, use, and price of fossil fuels were estimated at market-clearing quantities and prices; no waste discharge restrictions or capital availability limitations were imposed, and an unlimited supply of high-sulfur coal was assumed.

This base allocation, with no variations, was used in Cases 6.1 through 6.5 in the current environmental analysis. In other cases this base case allocation was varied to represent different levels of availability of clean fuels. Natural gas availability was reduced to the minimum amount necessary to operate emission-control equipment; low-sulfur coal availability was restricted to reflect the effects of strip-mining controls. Different assumptions for the electricity supply analysis, factor demand analysis, and capital demand analysis are specified in the respective chapters.

Capital Availabilities

Of the three types of resource constraints considered, capital is probably the most difficult to forecast because the supply and demand functions for capital are subject to so many external political and institutional variables.

In theory, capital, like any other resource, has a supply curve that is a function of price; in this case, the price of capital is the interest rate that must be paid to obtain it. By interest rate, the *real* rate of return is meant, or the nominal interest rate discounted by inflation. Broadly speaking, the higher the interest rate, the more capital will be made available at that rate. Thus, high rates in the U.S. relative to the rest of the world will attract additional capital to the U.S., as well as lure some additional capital out of hoarding. By the same token, the demand for capital decreases as the real interest rate increases; only those who are able to realize a rate of return on the borrowed capital at least as high as the interest rate they must pay will have an incentive to borrow. In theory, supply and demand balance at the equilibrium real interest rate.

In actuality, the political and institutional constraints on the free flow of capital are so complex as to make precise projections of the future supply of capital and interest rates extremely hazardous. Over the next decade there is the possibility that capital availabilities will fail to meet projected capital needs.

Capital spending by electric utilities was 11.8% of all capital spending by business in 1974; this percentage has been growing over the past decade (see Table 4.1). On the other hand, electric utilities' rate of return on investment has been declining and is now considerably below that of U.S. industry in general. The median rate of return on total capital for utilities was 5.8% in 1974; the median for all industry was 8.4% (*Forbes*, January 1975).[4]

Unless electric utilities can obtain enough rate relief from the regulatory commissions to offer a more competitive rate or return than they have in recent years, they may not be able to maintain their present share of the capital market, let alone increase it, particularly if interest rates remain high. Joskow and MacAvoy (1970)[5] estimate that regulatory commissions will have to raise targeted rates of return from the present 10-12% to 14-16% if utilities are to be able to meet projected capital needs at long-term interest rates of 10%.

To estimate the availability of new investment capital to electric utilities, two factors must be considered: the total capital available for new business investment and the percentage share of the total that utilities can obtain.

Table 4.1. Electric Utility Investment in Relation to Total Business Investment, 1960-1974

Year	Nonresidential Gross Domestic Private Investment ($ billion)	Electric Utilities Plant & Equipment Investment ($ billion)	Relative Importance of Electric Utilities %
1974	149*	17.6*	11.8
1973	137*	15.9*	11.6
1972	118	14.5	12.3
1971	104	12.9	12.4
1970	100	10.7	10.7
1969	98.5	8.94	9.1
1968	88.8	7.66	8.6
1967	83.3	6.75	8.1
1966	81.6	5.38	6.6
1965	71.3	4.43	6.2
1964	61.1	3.92	6.5
1963	54.3	3.67	6.8
1962	51.7	3.53	6.8
1961	47.0	3.55	7.6
1960	48.4	3.62	7.5

*Estimated from United States Department of Commerce, Social and Economic Statistics Administration, Bureau of Economic Analysis, *1973 Business Statistics: 19th Biennial Edition,*

An approximate estimate of future capital needs may be made by relating the total level of business investment to the gross national product (GNP). Historically, nonresidential investment has averaged 10.5% as a percentage of GNP; however, recent studies have projected capital needs at 11.4% to 12.3% of GNP over the next decade (Wallich, 1975).[6] This could imply substantial shortages, particularly if savings rates decline and large government deficits continue.

In this study, the average annual real growth rate over the period 1960-1967 will be used as a high estimate of GNP growth, namely, 4.2%/yr; the average real growth rate of 2.3%/yr for the period 1968-1972 will be used as the low estimate rate. These estimates, combined with the high (12.3%) and low (10.5%) estimates for business capitalization, and cumulated over the decade 1975-1985, result in estimates of capital needs (in billions of 1973 dollars) as shown in Table 4.2.

Table 4.2. Cumulated Business Capital Projections, 1975-1985
(billions of 1973 dollars)*

	Low Projections	High Projections
GNP growth	2.3%	4.2%
Business capital rate	10.5%	12.3%
Business capital supply	$1,682 billion	$2,213 billion

*Federal Energy Administration[1]

The electric utilities' ability to obtain investment capital in the capital markets in any given year will depend on such factors as their debt structure, their earnings prospects, and the prevailing long-term interest rate. All of these factors are dependent on government and regulatory agency decisions which cannot be forecast here. Instead, two rather simple alternative assumptions about electric utilities' market share for investment capital will be investigated. The first alternative is that electric utilities will maintain their present market share—11.8%. The second alternative is that their share declines to the average share realized for the period 1965-1974, or 9.7%. The possibility of the electric utilities' share increasing was not investigated in view of the foregoing discussion.

These market shares of 11.8% and 9.7% were then applied to the range of total business capital, as given in Table 4.2, to determine a probable range for investment capital available to electric utilities for the period 1975-1985. The high estimating procedure yields an estimate of $261 billion, the low procedure yields an estimate of $163 billion. Each of these figures is shown in Table 4.3. The question arises as to whether a set of estimates of environmen-

Table 4.3. Projected Cumulated Capital Availabilities for the Electric Utility Sector, 1975-1985

(billions of 1973 dollars)

	Low Projections	High Projections
Total capital	$1,682	$2,213
Share of market	9.7%	11.8%
Electric utility capital	$163	$261

tal constraints, fuel availabilities (both by type and total), and production requirements constitutes an achievable environment for the electric power industry, especially when coupled with possible limitations on capital availability. What degree of flexibility does the industry have regarding the substitution possibilities for different fuels, capital resources, and waste treatment alternatives? What will be the cost to the consumer in increased prices for meeting a specific set of criteria? Further, what will be the environmental consequences of using pollution taxes on the industry in a setting of limited capital? These questions are considered in Chapter 9.

References

1. Federal Energy Administration, *Project Independence Report: Project Independence*, U.S. Government Printing Office, Washington, D.C., 4118-00029, November 1974.
2. United States Congress, Senate Committee on Public Works, Hearings before the Subcommittee on Air and Water Pollution, 93rd Congress, *The Administration's Proposal for Relaxation of Air Pollution Standards*, 93-H19, US Government Printing Office, Washington, D.C., 1973.
3. Russell G. Thompson, R.J. Lievano, Robert R. Hill, James A. Calloway, and John C. Stone, "Relationship Between Supply/Demand and Pricing for Alternate Fuels in Texas: A Study in Elasticities," Final Report for the Texas Governor's Energy Advisory Council, Project S/D-4, Texas Office of the Governor, Office of Information Services, Austin, Texas, December 31, 1974.
4. "Utilities: Things Were Looking a Bit Better for the Utility Industry as 1974 Ended—But Not Better Enough," *Forbes*, V. 115, No. 1, pp. 144-152, January 1, 1975.
5. P.L. Joskow and P.W. MacAvoy, "Regulations and the Financial Condition of the Electric Power Companies in the 1970's," *American Economic Review*, V. 65, No. 2, pp. 293-301, 1970.
6. Henry Wallich, "Is There a Capital Shortage?," *Proceedings of the International Money Conference*, Amsterdam, June 11, 1975.

5. Electricity Use

Rodrigo J. Lievano ▪ James A. Calloway
Russell G. Thompson

Electricity Demand Projections

Prior to 1973, projections of the future use of electricity were of concern mainly to regulatory agencies such as the Federal Power Commission and to individual utilities. Electricity use had shown such regular growth in the previous 20-25 years [averaging 6.5%/yr (Dupree, 1972)][1] that projections could be made merely by extending the trend. Since then, major changes have taken place. In 1967, environmental considerations began to constrain capacity expansion decisions. The Clean Air Act Amendments of 1970 placed restrictions on pollutant emissions to the air, forcing utilities to either switch to costlier clean fuels or to install control devices. These restrictions placed an added burden on clean fuels such as natural gas, low-sulfur oil and low-sulfur coal, which were already relatively scarce. The OPEC oil embargo of late 1973 aggravated the situation to the extent that the Environmental Protection Agency (EPA) had to grant a large number of variances to utilities for the use of high-sulfur fuels (Council on Environmental Quality, 1974).[2] These developments greatly increased interest in projecting electricity demand, and a number of forecasting studies were performed by government agencies and other interested groups. Most of these studies were concerned with the entire energy supply and demand problem and not specifically with electric power generation, but the great importance of electric power generation as the largest and fastest growing single energy user [26% of total U.S. gross energy use and 22% of fossil fuel use in 1972 (Dupree, 1972)][1] required that particular attention be focused on this sector.

The newer studies abandoned trend projection methodologies in favor of methods which accounted for the response of electric power consumers to (1) increasing electricity prices and changes in personal income, (2) environmental restrictions on power plant siting and on fuel choice, and (3) possible changes in the growth rate of overall energy demand. Table 5.1 summarizes the findings of the more important studies in the order in which they were completed. This chronological ordering contrasts the pre-1973 studies based principally on trend projections with more recent ones which included the influences of the factors listed above. With the exception of the results shown for FORD-I (which reflect historical growth), all the later studies project a 1985 electricity demand significantly reduced from trend. The variations

Table 5.1. 1985 Electric Power Generation Projections

(10^{12} kWh)

Reference*		Total	Nuclear, Hydro, Other	Fossil Fuel
FPC	(1970)	4.44	1.80	2.64
NPC-I	(1972)	4.16	3.45	0.71
DW	(1972)	3.77	1.60	2.17
MCT-1	(1973)	3.70	†	†
MCT-2	(1973)	3.32	†	†
MCT-3	(1973)	2.55	†	†
FEA-7	(1974)	3.72	1.73	1.99
FEA-11	(1974)	3.62	1.73	1.89
FORD-1	(1975)	3.87	1.55	2.32
FORD-2	(1975)	2.40	0.82	1.58

*FPC —1970 National Power Survey—Federal Power Commission (1970)[3]
NPC-I —National Petroleum Council, Scenario I (1972)[4]
DW —Dupree and West (1972)[1]
MCT-1 —Mount, Chapman, and Tyrrell (1973)[5]—No change in real price
MCT-2 —Mount, Chapman, and Tyrrell (1973)[5]—Base case
MCT-3 —Mount, Chapman, and Tyrrell (1973)[5]—100% price increase
FEA-7 —Federal Energy Administration (1974)[6]—$7 oil, BAU Scenario
FEA-11 —Federal Energy Administration (1974)[6]—$11 oil, BAU Scenario
FORD-1 —Ford Foundation Energy Policy Project (1975)[7]—Historical Growth
FORD-2 —Ford Foundation Energy Policy Project (1975)[7]—Zero Energy Growth

†No projections or assumptions concerning fuel mix for generation

between the projections reflect different assumptions on electricity price, the influence of price on demand, personal income levels, the influence of income on demand, and population growth, as well as different methods of projection.

Price and Income Effects

The price of electricity and the personal income of consumers can affect electricity use in several ways. The demand for electricity is a derived demand resulting from consumer demand for a service which yields utility. These activities all require the use of equipment of relatively long life (heaters, stoves, lamps, etc.), and electricity provides the energy. As the price of electricity increases the consumer has several alternatives. In the short run, with a relatively fixed income, the consumer can curtail his use of electricity-consuming equipment by reducing comfort or lighting levels, etc. In the long run the consumer can respond by replacing worn-out equipment with new, more efficient equipment, by replacing it with equipment which utilizes another energy input, by substituting other inputs, such as improved insulation or natural lighting, for electricity, or by not replacing it at all. As his income rises, however, and the cost of electricity becomes a smaller portion of that income, the consumer may increase his demand for electric power through purchases of additional electrical appliances.

There is general agreement that the price elasticity of demand (the percent change in use relative to a percent change in price) and the income elasticity of demand for electricity (the percent change in use relative to a percent change in income) are much larger in the long run than in the short run in all use sectors (residential, commercial, industrial) (Taylor, 1975).[8] Table 5.2 lists the results of various econometric studies of electricity demand. These studies differ considerably in the data and methodolgy used, and considerable controversy exists as to whether the price variable used in most of the studies

Table 5.2. Estimates of the Price and Income Elasticities of Demand for Electricity

		Price Elasticity		Income Elasticity	
Reference	**Sector***	**Short-Run**	**Long-Run**	**Short-Run**	**Long-Run**
Fisher-Kaysen	R	−0.15	0	0.10	0
(1962)[9]	I	†	−1.25	†	†
Houthakker-Taylor	R	−0.13	−1.89	0.13	1.94
(1970)[10]					
Wilson					
(1971)[11]	R	†	−2.00	†	0
Halvorsen					
(1972)[12]	R	†	−1.14	†	0.52
Mount, Chapman,	R	−0.14	−1.20	0.02	0.20
Tyrell	C	−0.17	−1.36	0.11	0.86
(1973)[5]	I	−0.22	−1.82	†	†
Anderson					
(1973)[13]	R	†	−0.87	†	1.13
Levy	R	†	−1.11	†	0.44
(1973)[14]	(New England)				
Thompson et al.	R	−0.10	−0.67	0.28	1.87
(1975)[15]	(Texas)				
Griffin	R	−0.06	−0.52	0.06	0.88
(1974)[16]	I	−0.04	−0.51	0.89	0.89

*R: Residential
C: Commercial
I: Industrial

†Not given or not applicable

(average price) is really appropriate. The argument concludes that since electricity is sold by multi-step block pricing, in which the consumer price per kilowatt-hour of electricity decreases in steps as electricity use increases, the use of average price may lead to biased estimates of both the price and income elasticities. If electricity use is positively correlated with income levels, consumers at higher income levels will face lower rate schedules, and average price and income will be negatively correlated (Taylor, 1975).[8] The counter-argument is that most consumers (at least residential consumers) fall in a monthly consumption range in which the change in price (the marginal price) is very small, and that consumers do not have the knowledge of the utility's price structure required to make decisions based on marginal price (Anderson, 1973).[13] Additionally, it is argued, cross-sectional price data used in most studies contain more variation due to differences in electricity generation costs than to the rate structure. These arguments aside, the findings shown in Table 5.2 do indicate sensitivity of electricity demand to price and income levels. Some support for this is evidenced by the fact that total U.S. electrical output increased by only .48% in 1974—a period of increasing prices and low income growth (*Electrical World*, July 15, 1975).[17]

Nuclear and Hydro Development

The electric power industry model developed for this study is restricted to plants using fossil fuels (gas, oil, coal). Therefore, determining the extent to which nuclear generating capacity will provide future power needs in 1985 is a problem. Reliable projections of nuclear development are extremely difficult to make at this time for several reasons. First, the use of nuclear power to date has been so small that there is scant basis for making trend projections. Second, existing nuclear facilities continue to experience frequent operating problems so that their load factor is substantially below that of new fossil fuel plants (0.45 as opposed to 0.70). This may or may not continue to significantly affect the total share of electric power supplied from nuclear sources. Third, and most problematical, is the delay on licensing approval, construction, and operation of nuclear plants because of challenges on environmental grounds. Although these challenges may be overcome within the next decade by improved safety in plant operations and by increased public acceptance generally, the longer-range environmental problem of disposing of large quantities of nuclear wastes remains unsolved and could conceivably prevent many projections of greatly increased nuclear development from materializing.

Current plans for nuclear and fossil fuel generating plants suggest that about 25% of generating capacity will be nuclear by 1985 (*Electrical World*, March 15, 1975).[18] Table 5.3 presents a survey of recent projections of nuclear generation in 1985. These estimates vary nearly sixfold in absolute amounts and indicate that nuclear production would represent from 17 to 66% of the total production. For our study, we assume that nuclear generation represents from 20 to 45% of the total generation in 1985.

A survey of hydroelectric projections is given in Table 5.4. Hydroelectric capacity is resource limited; hence the absolute hydro capacity is projected to

Table 5.3. Projections of Nuclear Power Supply in 1985

	Ford (1)					
	Historical Growth	Zero Growth	Dupree-West (2)	National Petroleum Council (3)	Federation of American Scientists (4)	AEC (5)
Production (10^{12} kWh)	1.250	0.523	1.130	.749-2.928 (6)	0.87 (7)	—
Percent of total power production	32	21.7	27.3	17-66	25 (8)	24.5 (9)

(1) *A Time to Choose,* Energy Policy Project of the Ford Foundation, Ballinger, Cambridge, Mass., 1975.
(2) *Oil and Gas Imports Issues,* Hearings before the Committee on Interior and Insular Affairs, U.S. Senate, January 10, 11, 22, 1973.
(3) *U.S. Energy Outlook,* National Petroleum Council, Washington, D.C., Dec. 1972.
(4) *Federation of American Scientists Reports,* V. 28, No. 3, March 1975.
(5) As reported in *Electrical World,* March 15, 1975.
(6) From Table 5 of reference (3) above assuming 10,000 Btu/kWh heat rate.
(7) Load factor of 0.5 applied to projected 200,000 MW capacity.
(8), (9) Represents percent of total capacity.

Table 5.4. Projections of Hydroelectric Power Supply in 1985

	Ford (1)			
	Historical Growth	Zero Growth	Dupree-West (2)	National Petroleum Council (3)
Production (10^{12} kWh/yr.)	0.300	0.300	0.470	0.316
Percent of total power production	7.7	12.4	11.4	7.2

(1) *A Time to Choose,* Energy Policy Project of the Ford Foundation, Ballinger, Cambridge, Mass., 1975.
(2) *Oil and Gas Imports Issues,* Hearings before the Committee on Interior and Insular Affairs, U.S. Senate, January 10, 11, 22, 1973.
(3) *U.S. Energy Outlook,* National Petroleum Council, Washington, D.C., Dec. 1972.

be fairly constant by all studies. In this study the 0.3 x 10^{12} kWh/yr production projection of the Ford study was used.

Generation Mix

In the model the assumptions concerning nuclear and hydroelectric development have been outlined above. Relative use of nuclear and hydro power is determined exogenously and allowed to vary over a certain range. The model allocates the remaining power production (after nuclear and hydro are subtracted) between old and new plants and among the various production technologies considered—conventional coal, oil, gas, combined-cycle (CC), and coal gasification with combined-cycle (CGCC). In general, different generation mixes can be expected to result as environmental constraints and fuel availabilities and prices are varied. The model chooses that mix of production technologies which is most efficient (i.e., it satisfies the specified demand at the least total cost) subject to the constraints.

The base estimates for 1974 for the three types of conventional fossil-fueled plants are shown in Table 5.5. Also shown are the switch-over possibilities and new technologies allowed for in the model. The total production in 1985 from old plants (i.e., plants in place at the end of 1974) is assumed not to exceed the present production from these plants. In fact, some plants probably would be retired over this decade and the others would have reduced load factors as new plants come on stream. The minimum total amount of production from old plants in 1985 was estimated on the basis of trends in load factors and retirements to be 1.375 x 10^{12} kWh (see Table 5.5).

Table 5.5. Production Technologies and Limitations on Use

(trillions of kilowatt hours)

1974 (base estimate)	Switchover Limitations	1985 Use Limitations	
		Old	New (no limitation)
Conventional coal	none	Conv. coal $\leq$ 1.375	Conv. coal
Conventional oil	to coal ($\leq$ 0.2923)	Conv. oil $\leq$ 0.2923	Conv. oil
Conventional gas	to coal ($\leq$ 0.3204)	Conv. gas $\leq$ 0.3204	Conv. gas
Total	total ($\leq$ 0.6127)	Total ($\geq$ 1.375) ($\leq$ 1.444)	Comb.-cycle CGCC

Parametric Variations on Electricity Demand

No attempt was made in this study to refine the forecasts of electricity use or the proportions of nuclear, hydroelectric, and fossil fuel generation in 1985. Instead, it was desired to estimate a likely range of electricity requirements from fossil fuel sources in order to show how different electricity requirements in this range will affect the least-cost solution of the linear electric power model developed in this study. These cost analyses were made for different waste discharge restrictions and fossil fuel availabilties.

The method used for estimating this likely range consisted of starting with a base-case demand projected by the historical growth trend in electricity use and then adjusting this figure by assumptions on price elasticity, electricity price increases, income elasticity, increases in real income, and the contributions of nuclear and hydroelectric generation in 1985. The adjustment was made by applying the following functions:

$$DT = BD + PE(\Delta P) + IE(\Delta I) \quad (1)$$

$$DFF = DT - NH \quad (2)$$

where: DT = Total demand for electricity
BD = Base demand for electricity
PE = Price elasticity of demand (long-run)
ΔP = Percent change in price (to 1985)
IE = Income elasticity of demand (long-run)
ΔI = Percent change in income per capita
DFF = Electricity requirements from fossil-fuel sources
NH = Nuclear and hydroelectric generation

The base demand for electricity (BD) was established by projecting total generation in 1970 (1.53×10^{12} kWh; 82.4% fossil, 16.3% hydro, 1.3% nuclear by the historical growth trend of 6.5%/yr, giving a total of 2.93×10^{12} kWh in 1985 (Dupree, 1972).[1] Using equation (1), this value was adjusted with high and low estimates of long-run price elasticity (−2.0 and −0.87, respectively) and high and low estimates of long-run income elasticity (1.94 and 0.52, respectively) in conjunction with a 35% increase in the real price of electricity and a 25% increase in real income. Estimates of variations in the real price of electricity were obtained from previous solutions of the electric power model; estimates of changes in real income were determined as the difference between projections of the 1960-1967 growth rate (3.8%/yr) and the 1968-1972 growth rate (1.9%/yr) to 1985. The resulting four values for DT, obtained from all combinations of high and low estimates, were used in equation (2) along with high and low estimates of the percentage contribution of nuclear generation in 1985 (45% and 20%, respectively) and an assumed contribution of 0.3×10^{12} kWh from hydroelectric sources. Table 5.6 lists the resulting eight values from which the lower and upper estimates of electric

Table 5.6. Electricity Requirements from Fossil Fuel Sources in 1985
(10^{12} kWh)

Case		DT	NH		DFF
I	PE = −2.0	3.36	High	1.81	1.55
	IE = 0.5		Low	0.97	2.39
II	PE = −2.0	3.72	High	1.97	1.75
	IE = 1.94		Low	1.04	2.68
III	PE = −0.87	3.75	High	1.99	1.76
	IE = 0.5		Low	1.05	2.70
IV	PE = −0.87	4.11	High	2.15	1.96
	IE = 1.94		Low	1.12	2.99

power demand used in this analysis were chosen. To reflect the combination of the high demand case in Table 5.6 (4.11×10^{12} kWh) and the lowest nuclear and hydro contribution given in Table 5.1 (0.82×10^{12} kWh in FORD-2), the upper limit on projected electricity requirements from fossil power sources was set at 3.3×10^{12} kWh. The lower limit on projected electricity requirements from fossil fuel sources was set at 1.5×10^{12} kWh, which represents minimal growth in electric power generation from fossil fuel sources.

References

1. W.G Dupree and J.A. West, *United States Energy Through the Year 2000,* U.S. Department of the Interior, December 1972.
2. Council on Environmental Quality, *Fifth Annual Report,* U.S. Government Printing Office, Washington, D.C., December 1974.
3. Federal Power Commission, 1970 National Power Survey, December 1971.
4. National Petroleum Council, *U.S. Energy Outlook*, Summary Report of the National Petroleum Council, Washington, D.C., December 1972.
5. T.D. Mount, L.D. Champman, and T.J. Tyrrell, "Electricity Demand in the United States: An Econometric Analysis," Report No. ORNL-NSF-49, Oak Ridge National Laboratory, Oak Ridge, Tenn., June 1973.
6. Federal Energy Administration, *Project Independence*, Project Independence Report, November 1974.
7. Ford Foundation Energy Policy Project, *A Time to Choose: America's Energy Future*, Ballinger Publishing Co., Cambridge, Mass., 1975.
8. L.D. Taylor, "The Demand for Electricity: A Survey," *The Bell Journal of Economics*, American Telephone and Telegraph Co., New York, N.Y., Vol. No. 1, pp.74-110, Spring 1975.
9. F.M. Fisher and C.A. Kaysen, *A Study in Econometrics: The Demand for Electricity in the United States*, North-Holland Publishing Co., Amsterdam, The Netherlands, 1962.

10. H.S. Houthakker and L.D. Taylor, *Consumer Demand in the United States*, 2nd. ed., Harvard University Press, Cambridge, Mass., 1970.
11. J.W. Wilson, "Residential Demand for Electricity," *Quarterly Review of Economics and Business*, Vol. 11, No. 1, pp. 7-19, Spring 1971.
12. R. Halvorson, "Residential Electricity: Demand and Supply," paper presented at Sierra Club conference on Power and Public Policy, Vermont, January 1972.
13. K.P. Anderson, "Residential Energy Use: An Economic Analysis," Research Report No. R-1297-NSF, Rand Corporation, Santa Monica, California, October 1973.
14. P.F. Levy, "The Residential Demand for Electricity in New England," Research Report No. MIT-EL 73-017, Massachusetts Institute of Technology Energy Laboratory, November 1973.
15. R.G. Thompson, R.J. Lievano, R.R. Hill, J.A. Calloway and J.C. Stone, "Relationship Between Supply, Demand and Prices for Alternate Fuels in Texas: A Study in Elasticities," Final Report to the Texas Governor's Energy Advisory Council, December 31, 1974.
16. James M. Griffin, "The Effects of Higher Prices on Electricity Consumption," *The Bell Journal of Economics*, Vol. 5, No. 2, pp. 515-539, Autumn 1974.
17. *Electrical World*, "FPC Releases 1974 Statistics on Utilities," Vol. 184, No. 2, p. 28, July 15, 1975.
18. *Electrical World*, Annual Statistical Report, Vol. 183, No. 10, March 15, 1975.

6. Analysis of Environmental Restrictions

James A. Calloway • L. Ted Moore
Russell G. Thompson

With the linear electric power model structured to meet a national demand for electric power of 2.5 trillion kWh, a set of analyses was conducted to determine the effects of increasingly restrictive environmental control standards on production costs and production configuration in the electric power industry. Finally, an analysis was made to ascertain the extent to which waste treatment policies for wastewater discharges and air emissions may complement or impede one another.

Description of Analyses

Resource prices and fuel availabilties assumed for each anaylsis reflect 1985 projections and are summarized in Table 6.1. Projected fuel costs (in 1973 dollars) and availabilities are based on prices of $7.44/bbl for high-sulfur crude oil and 64¢/MSCF for natural gas at the well head. Low-, medium-, and high-sulfur coal are priced at $10.56/ton, $10.50/ton, and $10.37/ton, respectively. Fuel availabilities are based on projections obtained from a supply/demand/price study conducted for the Texas Governor's Energy Advisory Council (Thompson et al., 1974).[1] Projected demand for electric power generation from fossil-fuel sources is 2.5 trillion kWh.

Case 6.1 is the least restrictive of the five cases evaluated. This case allows the model to assume power production and waste treatment configurations

Table 6.1. Resource Price and Availability Assumptions

Resource	Price	Availability (10^{15} Btu)
Natural gas	64¢/MSCF	5.7
Low-sulfur oil	$7.88/bbl	0.2
High-sulfur oil	$7.44/bbl	1.5
Low-sulfur coal	$10.56/ton	6.16
Medium-sulfur coal	$10.50/ton	6.00
High-sulfur coal	$10.37/ton	Unlimited

which existed in 1974. The major waste treatment assumption involves use of particulate removal technology in all coal-burning and oil-burning plants.

Case 6.2 assumes that maximum restrictions of zero discharge are imposed on wastewater discharges. Use of particulate removal equipment is also assumed.

Case 6.3 places additional restrictions on waste discharges by assuming that existing (1974) standards apply to old plants and new source standards apply to new plants. All other parameters are the same as for Case 6.1.

Case 6.4 places maximum restrictions on air emissions. The model assumes use of particulate removal and stack gas scrubbing technology such that no untreated effluent is emitted.

Case 6.5 is the most restrictive case. It imposes maximum restrictions on both air and water discharges. It also imposes maximum restrictions on discharge of heat to water by disallowing use of the once-through cooling option.

EPA has promulgated 1977 "Best Practicable Technology" (BPT) standards and 1983 "Best Available Technology" (BAT) standards for wastewater pollutants. The corresponding air standards call for restricting both particulate and SO_2 effluents severely. Water effluent standards effective in 1977 restrict chemical pollutant discharges and increase the use of cooling towers; by mid-1983 no once-through cooling will be allowed (EPA, 1974).[2] New source standards on plants completed after July 1, 1977 require the use of cooling towers.

Case 6.1 is not in compliance with BPT standards for water pollutants. Case 6.2 is in compliance with a zero discharge water policy but not with a restrictive SO_2 policy. Case 6.3 requires new source standards on new plants only. Case 6.4 is only in partial compliance with 1983 BAT standards because not all old plants are retrofitted with cooling towers. Case 6.5 corresponds to zero discharge for water and maximum achievable constraints for controlling air pollutants.

Effect of Waste Treatment Restrictions

The general effects of increasingly severe waste discharge restrictions on electric power production costs are depicted in Figure 6.1. As each standard becomes increasingly strict, the cost of producing electric power continually increases from 1.27¢/kWh to 1.9¢/kWh (see Table 6.2).

Using the environmental constraints of Case 6.1 as the base case, the electric power generation mix consists of combined-cycle plants and conventional plants burning high-sulfur coal. Old high-sulfur coal-burning plants with particulate removal meet 50% of the electricity requirements; new high-sulfur coal-burning plants with particulate removal meet 19%; and new combined cycle plants provide 27%. All natural gas available for electric power

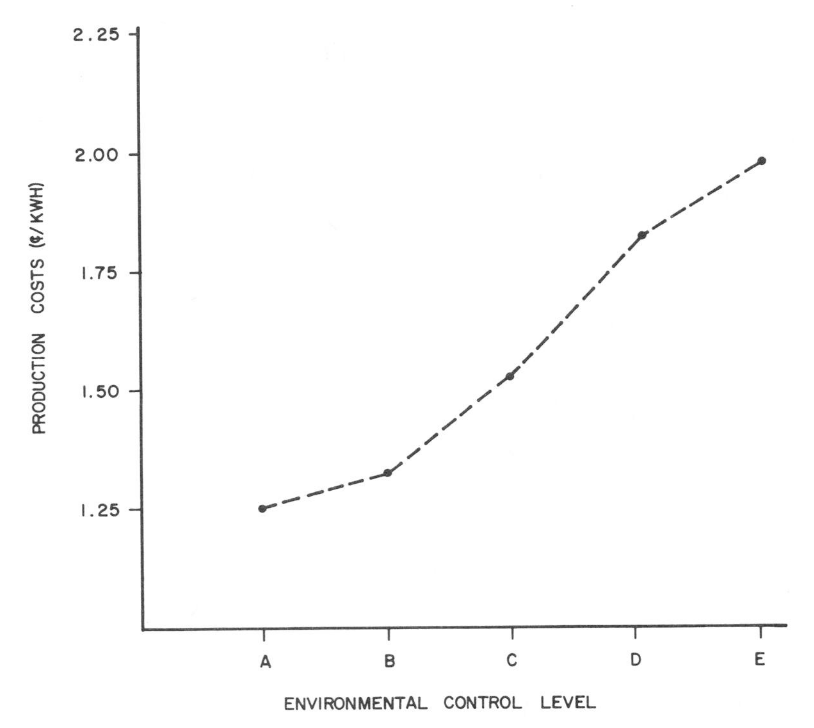

Figure 6.1. Production costs vs. environmental restrictions in steam electric power generation.

generation is used to supply the high efficient combined-cycle plants. With oil priced at $7.44/bbl and environmental restrictions at a minimum, high-sulfur coal is the most economical fuel after natural gas; the model reflects a complete shift away from oil- and gas-fired boilers to the use of combined-cycle plants and high-sulfur coal-burning plants.

The production mix and production costs for each of the cases are summarized in Table 6.2. Resource use and the marginal value of each of the major resources are tabulated in Tables 6.3 and 6.4, respectively. Table 6.5 sum-

Table 6.2. Production Generation Mix for Five Cases of Increasing Environmental Control

Plant Type	Cases 6.1	6.2	6.3	6.4	6.5
	Percent Generation Mix				
Old Plants					
Low-sulfur oil	—	—	0.7%	0.7%	0.09%
High-sulfur oil	—	—	5.3	5.2	5.2
Low-sulfur coal	—	—	19.7	22.8	22.7
Medium-sulfur coal	—	—	22.5	19.6	20.1
High-sulfur coal	50%	49%	0.8	—	—
Combined-cycle	4	4	3.8	3.8	3.7
Total old plants	54	53	52.8	52.1	51.8
New Plants					
Low-sulfur coal	—	—	4	—	—
Medium-sulfur coal	—	—	—	3.1	2.3
High-sulfur coal	19	20	17.6	21.2	—
Combined-cycle	27	27	25.6	23.6	25.0
Coal gasification combined-cycle	—	—	—	—	20.9
Total new plants	46	47	47.2	47.9	48.2
Total power production for in-plant use and sale (trillion kWh)	2.555	2.584	2.592	2.632	2.657
Production costs (¢/kWh)	1.27	1.35	1.57	1.82	1.90

marizes water use and waste discharges for each case. Coal-burning plants are assumed to use particulate removal equipment; thus, for Case 6.1, particulate discharges are low (0.499 lb./MMBtu). Unrestrained sulfur dioxide discharges amount to 13 lbs./MMBtu of coal consumed. Average stack heat discharges were 1000 Btu/kWh, while heat discharged to water was 3650 Btu/kWh. Waste discharges to water were dissolved solids (5.10 lbs./kWh) and suspended solids (0.000132 lb./kWh). Water withdrawals for the base case were 28.78 gal./kWh. Since a high degree of once-through cooling was used, actual in-plant water consumption was a small fraction of withdrawals (0.105 gal./kWh).

Investment distribution for all five cases is given in Table 6.6. All investment figures are in constant 1973 dollars. Total investment in new plants and equipment for the 10-year period 1975-1985 is $40.09 billion in Case 7.1. It is

Table 6.3. Resource Use to Produce 2.5 Trillion Kilowatt Hours of Electric Power

(10^{12} MBtu)

	Cases				
	6.1	**6.2**	**6.3**	**6.4**	**6.5**
Natural gas	5.7	5.7	5.7	5.7	5.7
Low-sulfur coal	0	0	6.16	6.16	6.16
Medium-sulfur coal	0	0	6.0	6.0	6.0
High-sulfur coal	17.2	17.45	4.137	4.788	4.58
Low-sulfur oil	0	0	0.2	0.2	0.2
High-sulfur oil	0	0	1.5	1.5	1.5
Total fuel	22.9	23.2	23.7	24.4	24.1
Heat rate (Btu/kWh)	9160	9280	9480	9760	9640

Table 6.4. Marginal Values in Electric Power Generation for Environmental Cases 3, 4, and 5

($/MM Btu)

	Cases		
	6.3	**6.4**	**6.5**
Natural gas	2.31	2.08	2.10
Low-sulfur coal	1.17	1.14	1.10
Medium-sulfur coal	0.861	0.758	0.712
Low-sulfur oil	0.530	0.445	0.443
High-sulfur oil	0.545	0.224	0.220

distributed approximately equally between construction of combined-cycle (52.4%) and high-sulfur coal-burning plants (45.9%). Investment, as well as production cost, increases substantially as discharge constraints become more severe. Again, it must be recalled that capital costs include neither escalation nor interest during construction.

Case 6.2 imposes zero discharge restrictions on waterborne pollutants, causing a very slight shift in the generation mix. Total power production of the industry increases approximately 1% to meet the power requirements of wastewater treatment equipment. Old high-sulfur coal-burning plants produce 49% of the national requirements, new high-sulfur coal-burning plants 20%, and new combined cycle plants 27%. Total new investment becomes $43.45 billion, again divided almost equally between combined-cycle (48.3%) and coal-burning plants (44.9%).

Except for discharges of dissolved and suspended solids, which were constrained to zero, waste discharges varied little from the base case. The most significant effect was the reduction of water withdrawals of 98% from 28.78

Table 6.5. Water Use and Waste Discharges in Electric Power Generation for Cases 1, 2, 3, 4, 5 of Environmental Restrictions Analysis

	Cases				
	6.1	6.2	6.3	6.4	6.5
Water (gal./kWh)					
Withdrawals	28.78	0.55	20.37	30.34	0.589
In-plant consumption	0.105	0.55	0.314	0.216	0.589
Heat (MBtu/kWh)					
Stack	1.00	1.01	1.04	1.06	0.84
Water	3.65	0	2.55	3.84	0
Water Pollutants (lbs./kWh)					
Dissolved solids	5.10	0	3.69	5.34	0
Suspended solids	0.000132	0	0	0.000132	0
Air Pollutants (lbs./MMBtu)					
Sulfur dioxide from coal	13.0	13.0	3.19	0.61	1.07
Particulates from coal	0.499	0.499	0.498	0.496	0.376

Table 6.6. Investment Distribution

	Cases				
Plant Type	6.1	6.2	6.3	6.4	6.5
	Percent Distribution				
Low-sulfur coal	—	—	8.5%	—	—
Medium-sulfur coal	—	—	—	4.5%	3.5%
High-sulfur coal	45.9%	44.9%	36.2	31.2	—
Low-sulfur oil	—	—	Trace	—	—
Combined-cycle	52.4	48.3	41.3	27.3	29.4
Coal gasification combined-cycle	—	—	—	—	25.3
Other*	1.7	6.8	14	37	41.8
Total investment ($/billion)	40.09	43.45	48.17	68.25	67.89

*Primarily waste control investment

gal./kWh to 0.55 gal./kWh, through complete substitution of wet-cycle cooling towers for once-through cooling. This use of cooling towers has the added effect of eliminating most of the heat discharges to water.

The cost of producing electric power increased from 1.27¢/kWh to 1.35¢/kWh in applying the restrictions of Case 6.2. This represents a 6% increase for meeting zero water discharge standards. Total energy use in the overall industry increased from 9160 Btu/kWh to 9280 Btu/kWh from Case 6.1 to 6.2, an increase of approximately 1.3% (see Table 6.3).

The restrictions imposed in Case 6.3 cause the industry to retain a substantial portion of old oil-burning plants using both high- and low-sulfur oil. Additionally, there is a small amount of investment in new low-sulfur oil-burning plants. There is an approximate 1½% increase in total electric power production over that in the base case; the generation mix consists of high-sulfur oil (5.3%); low-sulfur oil (0.7%); old plants using high-sulfur coal (0.8%), medium-sulfur coal (22.5%), and low-sulfur coal (19.7%); new plants using high-sulfur coal (17.6%) and low-sulfur coal (4%); and new combined-cycle plants (25.6%). Total investment increased slightly to $48.17 billion and was distributed among combined-cycle (41.3%), high-sulfur coal (36.2%), and low-sulfur coal (8.5%). In old coal-burning plants where existing standards can be met without making investment in expensive stack gas scrubbing equipment the shift is to low- and medium-sulfur coals. In new coal-burning plants there is a substantial but smaller shift to low-sulfur coal as well.

The waste discharge standards imposed in Case 6.3 represent existing (1974) standards for existing plants and new source standards for new plants. These standards result in a slight decrease in the use of once-through cooling and an accompanying loss in plant efficiency. Total energy use increased approximately 3½% from the base case to 9840 Btu/kWh, and production costs increased 23.6% to 1.57¢/kWh.

Water withdrawals decrease slightly from the base case (20.37 gals./kWh) with a subsequent decrease in dissolved solids discharged (3.69 lbs./kWh) Sulfur dioxide emissions were reduced significantly to 3.19 lbs./MMBtu o coal consumed because of new source standards on new production equip ment.

Case. 6.4 imposes the additional restriction of allowing no untreated ai emissions. In the model the stringency of this requirement cannot be met b use of low-sulfur fuels alone, so stack gas scrubbers are required. Substantia new investment is required to meet this constraint, and the generation mi shifts again. Power production increases another 1.5% from Case 6.3, an high-sulfur coal-burning plants are replaced by increased use of low-sulfu coal-burning plants. The generation mix consists of old high-sulfur (5.2% and low-sulfur (0.7%) oil plants; old medium-sulfur (19.6%) and low-sulfu (22.8%) coal plants; new high-sulfur (21.2%) and medium-sulfur (3.1%) co plants; and new combined-cycle plants (25.6%).

Total investment increased substantially to $68.25 billion. Sixty-three percent of the total investment was used for new production capacity: combined-cycle (27.3%), high-sulfur coal (31.2%) and medium-sulfur coal (4.5%).

Under the imposed restrictions of no untreated air emissions, sulfur dioxide emissions were reduced well below new source standards for the entire industry. Total SO_2 emissions were 0.61 lbs./MMBtu of coal consumed. Energy use increased still further to 9760 Btu/kWh resulting primarily from increased use of stack gas cleaning equipment. This is an increase of 6.5% from the base case. Production costs increased 1.82¢/kWh, an increase of 43.3% over the base case.

The most restrictive case, Case 6.5, imposes zero discharge of water pollutants (including heat) and allows no discharge of untreated air emissions. Because burning of high-sulfur coal contributes to water pollution problems as well as air pollution problems, the major shift in generation mix for this case involved elimination of high-sulfur coal as a direct fuel for steam boilers. The generation capacity lost was directly replaced by coal gasification combined-cycle (CGCC) technology using high-sulfur coal. The generation mix for Case 6.5 is virtually unchanged from Case 6.4 except for the substitution of CGCC plants for high-sulfur coal-burning plants. Total power production increased by 4% over Case 6.1 to meet in-plant demands. Investment decreased slightly to $67.89 billion. Sixty percent of the investment was for new production capacity: combined-cycle (29.4%), coal gasification combined-cycle (25.3%), and medium-sulfur coal (3.5%) plants.

Case 6.5 requires zero discharge of water pollutants, maximum treatment of air emissions, and use of wet-cycle cooling towers. Because of the shift in technology, namely to coal gasification in lieu of particulate and SO_2 removal equipment, air emissions vary slightly from Case 6.4. SO_2 emissions increase slightly to 1.07 lbs./MMBtu of coal consumed, and particulate emissions are reduced to 0.376 lb./MMBtu. Increases in SO_2 emissions are attributed to somewhat less efficient sulfur removal in coal gasification. Energy use increases approximately 5% to 9640 Btu/kWh from the base case but decreased slightly from Case 6.4 because of increased use of the more efficient gas-burning combined-cycle plants. Combined-cycle plants require less cooling than conventional plants and need no particulate cleanup. Production costs increase by 49.6% from the base case to 1.9%/kWh.

In summary, all available natural gas is used as fuel for combined-cycle plants or for gas reheating in stack gas cleaning equipment. Increasing the controls on water and air discharges causes a shift away from use of high-sulfur coal as boiler fuel in steam electric plants. Under the most severe restriction, CGCC plants using high-sulfur coal become economically feasible. Total investment required to meet both projected power demands of 2.5 trillion kWh and environmental restrictions for 1985 is approximately $68 billion. In each case approximately $20 billion of new investment is allocated for new combined-cycle plants. *The conclusion drawn is that simultaneously meeting projected 1985 demands for electric power and the 1985 environmental standards is likely to cause capital cost increases of 50% or more.*

Complementarity Analysis

In many cases, imposition of restrictive air emission controls increases the cost of wastewater cleanup. The use of control technology converts air pollution problems into water pollution problems or vice versa. A number of studies have been conducted to determine the projected costs of a specific level of air pollution control or water pollution control. Given this possible complementary effect, the question arises as to whether the costs obtained from separate analyses may reasonably be added together to obtain a meaningful estimate of the total cost of treating both air and water pollution problems. Cases 6.1, 6.2, 6.4 and 6.5 were used to analyze the complementary effect on electric power generation except that particulate removal as eliminated from Cases 6.1 and 6.2 to obtain strictly singular treatment costs.

Briefly, Case 6.1, as modified above, is the base case having minimum environmental restrictions, and Case 6.2 employs restrictions on water discharges only. Case 6.4 controls air emissions only, and Case 6.5 places maximum restrictions on both air and water discharges. A complementarity coefficient is determined by the relationship

$$K = \frac{(C_2 - C_1) + (C_4 - C_1)}{(C_5 - C_1)} \quad (1)$$

The numerator simply expresses the sum of the differential water pollution control costs and air pollution control costs taken separately; the differential costs of both air and water control taken together are expressed in the denominator.

If the factor K is greater than one, the sum of the parts is greater than the whole. That is, the cost of treating air emissions plus the cost of treating water emissions is greater than the cost of treating both simultaneously. In such an event, air and water treatment processes enhance one another. If the ratio is less than one, air and water treatment processes reduce the effectiveness of one another.

The total cost of meeting projected 1985 demand for electric power under each of the cases listed is given below:

Scenario	*Production Costs ($ billion)*
Base case	26.420
Cost using water treatment only	27.717
Cost using air treatment only	45.398
Cost using both air and water treatment	47.561

Using equation (1), the estimate of the complementarity coefficient is 0.958. This result indicates that separate estimates of air pollution control and water pollution control, when added together, underestimate costs of clean water and clean air in electric power generation by 4.2%. However, the degree of underestimation seems relatively small. One must therefore conclude that the complementarity effect for electric power is not a factor and that estimates of costs taken separately may be added to estimate total costs of air and water cleanup.

References

1. R.G. Thompson, R.J. Lievano, R.R. Hill, J.A. Calloway and J.C. Stone, "Relationship Between Supply, Demand and Prices for Alternate Fuels in Texas: A Study in Elasticities," Final Report to the Texas Governor's Energy Advisory Council, December 31, 1974.
2. *Development Document for Proposed Effluent Limitations Guidelines and New Source Performance Standards for the Steam Electric Power Generating Point Source Category*—Environmental Protection Agency, 440/1-73/029, Washington, pp. 4-5, March, 1974.

7. Electricity Supply Analysis

Russell G. Thompson • James A. Calloway
L. Ted Moore

The electricity supply analysis evaluates the effects of different fossil fuel availabilities and waste discharge standards on the costs of producing electricity, the marginal value of resources and the marginal cost of waste discharge control for electricity requirements varying from 1.5 to 3.3 trillion kWh. The electricity supply analysis is made for the three combinations of fuel availabilities and waste discharge restrictions given in Table 7.1; fuel prices are given in Table 6.1.

In Case 7.1 considerable natural gas, fuel oil, and low-sulfur coal are assumed to be available. Waste discharge restrictions on air and water pollutants are similar to those being enforced today. In Case 7.2 natural gas is limited to the amount required for use in stack gas scrubbers; fuel oil and coal availabilities are the same as in Case 7.1. Existing waste discharge restrictions are assumed for old plants; new source standards including wet-cooling towers are assumed for newly-built plants. In Case 7.3 natural gas is again limited to the amount required for stack gas scrubbers, no fuel oil is available, and low-sulfur coal is limited to the amount available without new strip mining. Stack gas scrubbing of sulfur dioxide emissions and particulate control are required. Recycle cooling towers are required to dissipate the heat in the cooling water; no wastewater pollutants from the cooling tower blowdown may be discharged to the waterways.

Marginal Costs of Producing Electricity

The marginal costs of producing electricity are summarized in Figure 7.1 for the range of electricity requirements evaluated. Marginal costs are considerably lower in Case 7.1 than in Cases 7.2 and 7.3. Marginal costs are higher in Case 7.2, with more moderate waste discharge restrictions and clean fuel availabilities than in Case 7.3, which specifies stringent waste discharge restrictions and an extreme scarcity of clean fuels.

Marginal costs in Case 7.1 are low because natural gas is relatively abundant and high-sulfur coal can be used without penalty to produce electricity in old and new plants. Marginal costs in Case 7.3 are lower than those in Case 7.2 because the extreme scarcity of clean fuels and the stringent waste discharge restrictions in Case 7.3 make it economical to invest in the more efficient coal gasification combined-cycle plants; that is, the model substitutes capital costs for non-capital costs. The non-capital-to-capital substitution

Table 7.1. Specifications of Fuel Availabilities and Waste Discharge Standards for Cases 7.1, 7.2 and 7.3 of Electricity Supply Analysis

Case 7.1		Case 7.2		Case 7.3	
1. Fuel Availabilities (10^{12} MBtu)		**1. Fuel Availabilities (10^{12} MBtu)**		**1. Fuel Availabilities (10^{12} MBtu)**	
Natural gas	5.7	Natural gas	1.3	Natural gas	1.3
Low-sulfur coal	6.16	Low-sulfur coal	6.16	Low-sulfur coal	3.73
Medium-sulfur coal	6.00	Medium-sulfur coal	6.00	Medium-sulfur coal	6.00
High-sulfur coal	Unlimited	High-sulfur coal	Unlimited	High-sulfur coal	Unlimited
Low-sulfur oil	0.2	Low-sulfur oil	0.2	Low-sulfur oil	Zero
High-sulfur oil	1.5	High-sulfur oil	1.5	High-sulfur oil	Zero
2. Water Standards		**2. Water Standards**		**2. Water Standards**	
1974 proportion of once-through cooling		Existing cooling towers on existing plants		Zero discharge of dissolved solids and suspended solids	
		Cooling towers on new plants		Cooling towers	
3. Air Standards		**3. Air Standards**		**3. Air Standards**	
Particulate control		Existing standards on existing plants;		Particulate control	
No sulfur dioxide control		New source standards on new plants		Sulfur dioxide control	

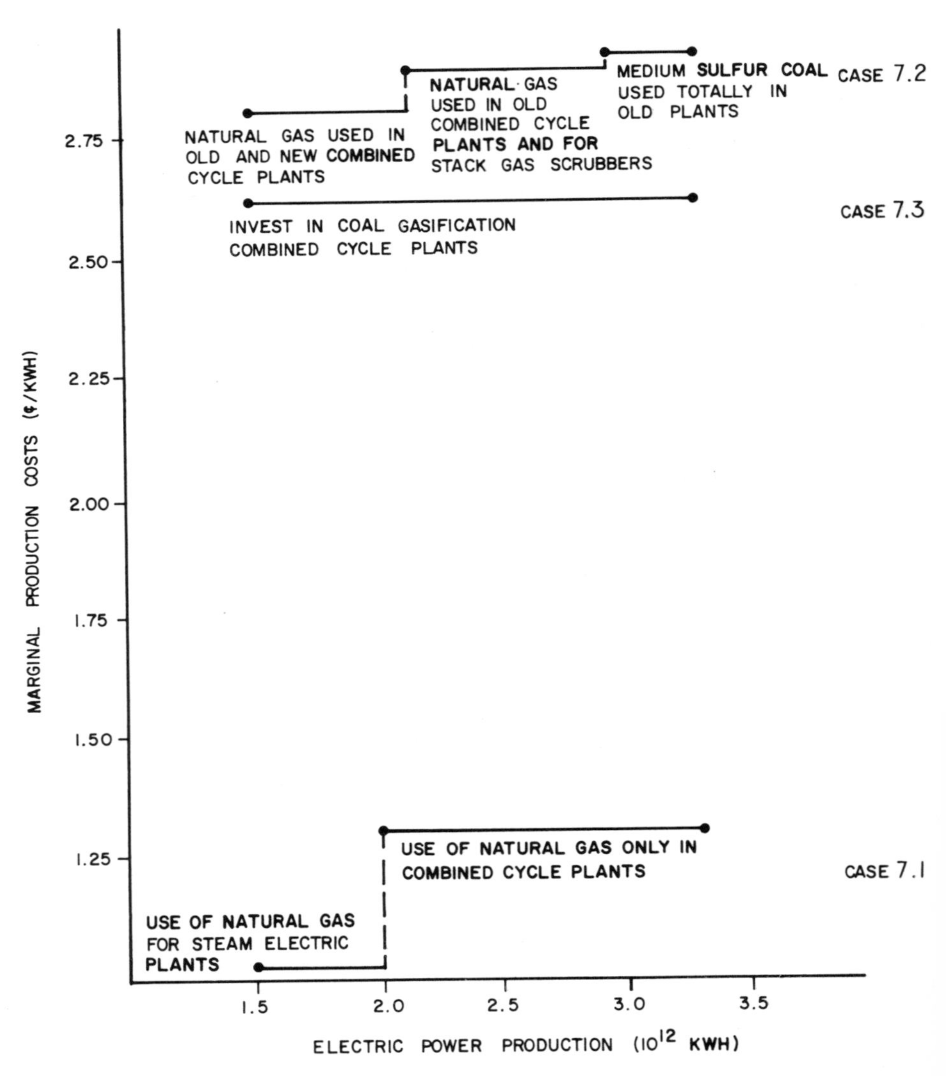

Figure 7.1. Marginal costs of electricity production for three different fossil fuel availabilities and environmental policy specifications.

ratio is 2.25 for Case 7.2 and 2.0 for Case 7.3. This substitution results in considerably higher average production costs in Case 7.3 than in Case 7.2 as shown below.

Investments in new plants and waste discharge control equipment in Case 7.2 totaled $35.23 billion at a production level of 2.17 trillion kWh and $69.31

billion at 2.96 trillion kWh. Comparable investments for Case 7.3 totaled $59.11 billion and $87.55 billion at these same levels of electricity requirements.

Capital availability limitations were analyzed in Cases 7.2 and 7.3. In Case 7.2 the model has considerable investment flexibility for all levels of electricity requirements. In Case 7.3, however, the severe waste discharge restrictions and limited fuel supplies severely restrict the investment substitution possibilities; in fact, in Case 7.3, no reduction in capital availability is possible.

Average Costs of Producing Electricity

The average cost curves for producing electricity are shown in Figure 7.2 for the range of electricity requirements evaluated. Average costs of producing electricity are lowest in Case 7.1, higher in Case 7.2, and highest in Case 7.3. With the continued availability of natural gas to fire steam boilers, electricity requirements in Case 7.1 are produced at decreasing average costs. However, when natural gas becomes scarce, at a production level of 2.03 trillion kWh, average costs increase gradually over the remaining range of electricity requirements. For Case 7.1, average costs of producing electricity are 1.31, 1.26, and 1.28¢/kWh at 1.5, 2.03, and 3.3 trillion kWh, respectively.

In Case 7.2, electricity requirements are satisfied at increasing average costs over the range of electricity requirements evaluated. At a production level of 1.5 trillion kWh, old plants, meeting existing waste discharge standards through heavy use of clean fuels, produce 89% of the electric power requirements. As the supply of clean fuels is exhausted, average costs increase continually thereafter at a decreasing rate and increased investments are made in new high-sulfur coal plants. These costs increase because an increasingly larger portion of the operating and capital costs is directed to meeting the new source standards for new high-sulfur coal plants. Average costs are 1.89, 2.08, and 2.14¢/kWh at production levels of 2.17, 2.96, and 3.3 trillion kWh, respectively.

In Case 7.3 electricity requirements are produced at constantly increasing average costs over almost the entire range of electricity requirements evaluated. Average costs are 2.35¢/kWh at the lowest range evaluated and 2.51¢/kWh at the highest range.

Marginal Values of Scarce Fuels

For all cases, the marginal values of clean fuels increase with increasingly restrictive environmental standards and increased requirements for electric power.

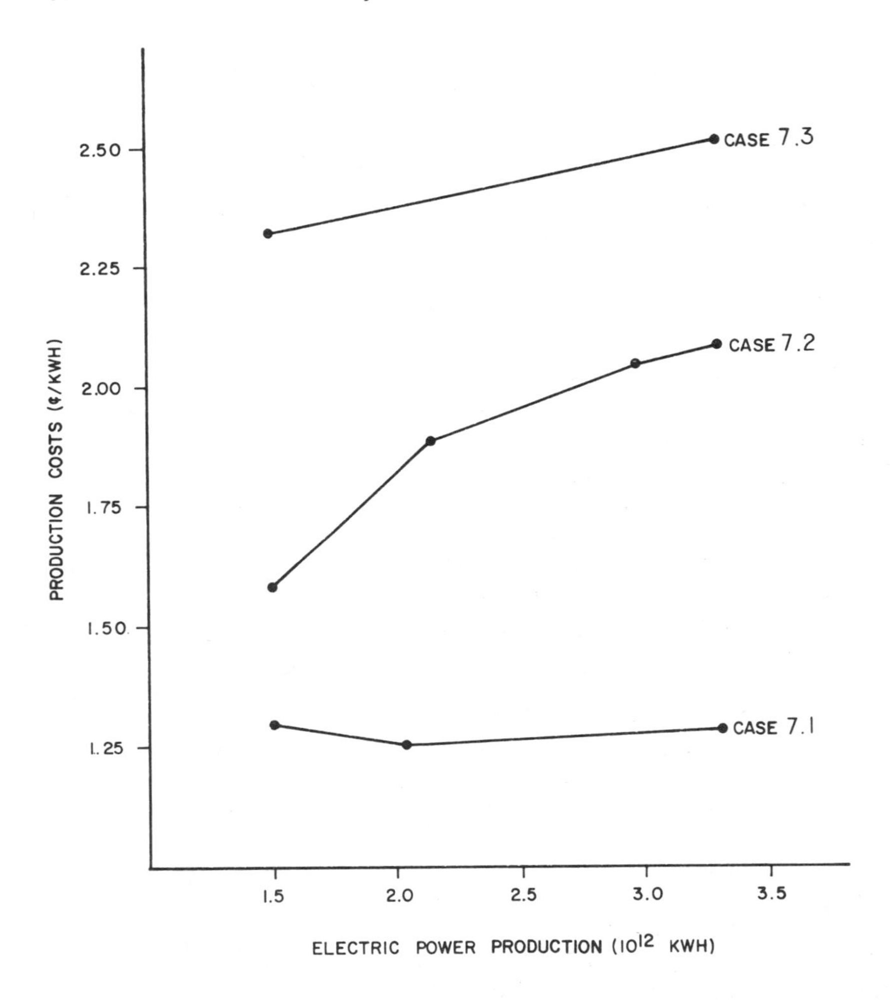

Figure 7.2. Average costs of electricity production for three different fossil fuel availabilities and environmental policy specifications.

In Case 7.1 natural gas is the only fuel which becomes scarce; its marginal value is 39.3¢/MMBtu. In Case 7.2 natural gas is scarce at all levels of electricity requirements; its marginal value increases from $2.38 to $2.79, and then to $3.19/MMBtu for increasing production requirements in the low, intermediate and high ranges. The marginal value of low-sulfur coal also increases, but at a much slower rate. A complete summary of the marginal values of scarce fuels is included in Table 7.2. In Case 7.3 the marginal values

Table 7.2. Marginal Values ($/MMBtu) of Fuels for Cases 7.1, 7.2, and 7.3 of Electricity Supply Analysis for Different Ranges of Electricity Requirements

	Case 7.1		Case 7.2			Case 7.3	
Fuels	Low	High	Low	Intermediate	High	Low	High
Natural gas	0	0.393	2.38	2.79	3.19	2.93	2.94
Low-sulfur coal	0	0	1.35	1.38	1.68	1.47	1.47
Medium-sulfur coal	0	0	0.89	0.91	1.21	1.06	1.07
High-sulfur coal	0	0	0	0	0	0	0
Low-sulfur oil	0	0	0.73	0.76	1.05	2.18	2.19
High-sulfur oil	0	0	0.69	0.71	1.00	2.15	2.16

of the scarce fuels are constant over almost the entire range of requirements from 1.5 to 3.3 trillion kWh, with the generation mix changing slightly at 3.26 trillion kWh. Natural gas has a marginal value of $2.93/MMBtu below 3.26 trillion kWh, increasing to $2.94 for the higher demand level. Similarly, low-sulfur oil's marginal value changes from $2.18 to $2.19/MMBtu, and the marginal value for high-sulfur oil changes from $2.15 to $2.16/MMBtu.

The production mix selected by the model to satisfy conditions reflected in each of the scenarios is given in Table 7.3. The generation technology selected for a given case is indicated by an asterisk (*).

Marginal Costs of Waste Discharge Control

The linear electric power model provides estimates of the marginal costs of waste discharge controls for sulfur dioxide, particulates, suspended solids, and dissolved solids. In Case 7.1 the only binding restriction is on particulate control. The marginal cost of particulate control is $1.81/lb. In Case 7.2 the only binding restrictions are on sulfur dioxide and particulates. Marginal costs of removing sulfur dioxide are $0.154/lb. in the low range, $0.156/lb. in the intermediate range, and $0.159/lb. in the upper range of electricity requirements. Marginal costs of removing particulates are $1.81/lb. in all three ranges of electricity requirements. In Case 7.3 the marginal costs of controlling particulates, sulfur dioxide, and dissolved solids are $1.44, $0.14, and $0.018/lb., respectively.

Effects of Water Quality Variation

There are large variations in water temperature and water quality in the U.S., and, since these factors affect the operational efficiency of electric

Table 7.3. Production Mix of Fuels for Cases 7.1, 7.2, and 7.3—Electricity Supply Analyses for Different Ranges of Electricity Requirements†

	Case 7.1		Case 7.2			Case 7.3	
Steam Boilers	**Low Range**	**High Range**	**Low Range**	**Intermediate Range**	**High Range**	**Low Range**	**High Range**
Natural gas, old plant	*						
Low-sulfur coal, old plant			*	*	*	*	*
Low-sulfur coal, new plant							
Medium-sulfur coal, old plant			*	*	*	*	*
Medium-sulfur coal, new plant			*	*			
High-sulfur coal, old plant	*	*				*	*
High-sulfur coal, new plant		*	*	*	*		*
Low-sulfur oil, old plant			*	*	*		
High-sulfur oil, old plant			*	*	*		
Combined-cycle							
Comb.-cycle, old plant	*	*	*	*	*	*	*
Comb.-cycle, new plant	*	*	*				
Coal Gasification com.-cycle							
CGCC, new plant						*	*

†The range of electricity requirements coincides with the range of constant marginal costs in Figure 7.1.
*An asterisk indicates generation method used.

power plants, the model was used to analyze the effects of water quality variations. Case 7.1 employed relatively relaxed air and water standards and Case 7.3 used the most stringent standards. In both cases, improving the quality of the input water reduced the cost of producing electric power (see Table 3.1 for base case water quality specifications). Using high quality (100 mg/liter of total dissolved solids) water decreased production costs by 2.68% in Case 7.1 and 3.06% in Case 7.3; using low quality (800 mg/liter of total dissolved solids) water increased costs less than 3% in both cases.

8. Derived Demand Analyses for Water Withdrawals and Low-Sulfur Coal

Russell G. Thompson ▪ James A. Calloway
L. Ted Moore

The derived demand analyses for water withdrawals and low-sulfur coal are used to show the effects of higher prices of these two important inputs to electric power generation on the uses of these inputs in the production of electricity. These evaluations are made for different assumptions with regard to electricity requirements, waste discharge restrictions, and fuel availabilities.

As in Chapter 7, no limitations are placed on the amount of capital available for investment or on water withdrawals available for use. Fuel prices, except for low-sulfur coal, are the same as specified in Table 6.1.

Derived Demand for Water Withdrawals

Three different analyses were conducted to determine the effect of water withdrawal price on the use of water withdrawals in electric power generation. The separate cases are described as follows: (1) low demand for electric power, lenient discharge restrictions, and relatively abundant fuels; (2) medium demand for electric power, relatively severe discharge restrictions, and the same fuel availabilities as in Case (1) except for natural gas; and (3) high demand, severe discharge restrictions, and reduced availability of low-sulfur coal with a price of $2.11/MMBtu. See Table 8.1 for a complete specification of these three cases.

Using a water withdrawal price of zero, a solution to the model was obtained for each scenario for the respective set of fuel prices, fuel availabilities, and discharge restrictions specified in Table 8.1. Holding all other parameters constant, the price of water withdrawals was then systematically increased to as much as $1/gal. to determine the effect of such increases on water withdrawals and plant configuration.

Case 8.1 of the water price analysis sets the minimum requirements for the analysis. Electric power demand for 1985 is set at the lowest level (1.5 trillion kWh) and discharge restrictions are set at the minimum (equal to Case 6.1) specified in the environmental restrictions section of Chapter 6. Fuel

Table 8.1. Specifications of Fuel, Waste Discharge, and Electricity Requirements for Cases 8.1, 8.2, and 8.3—Water Demand Analyses

Case 8.1		Case 8.2		Case 8.3	
1. Fuel Availabilities (10^{12} MBtu)		**1. Fuel Availabilities (10^{12} MBtu)**		**1. Fuel Availabilities (10^{12} MBtu)**	
Natural gas	5.7	Natural gas	1.3	Natural gas	1.3
Low-sulfur coal	6.16	Low-sulfur coal	6.16	Low-sulfur coal	3.88
Medium-sulfur coal	6.0	Medium-sulfur coal	6.0	Medium-sulfur coal	6.0
High-sulfur coal	Unlimited	High-sulfur coal	Unlimited	High-sulfur coal	Unlimited
Low-sulfur oil	0.2	Low-sulfur oil	0.2	Low-sulfur oil	Zero
High-sulfur oil	1.5	High-sulfur oil	1.5	High-sulfur oil	Zero
2. Water Standards		**2. Water Standards**		**2. Water Standards**	
1974 proportions of once-through cooling		1974 proportions of once-through cooling		Cooling towers; zero discharge of pollutants	
3. Air Standards		**3. Air Standards**		**3. Air Standards**	
No untreated particulates; no stack gas scrubbers		Existing standards for old plants; new source standards for new plants		Particulate control; stack gas scrubbers	
4. Electricity Requirements		**4. Electricity Requirements**		**4. Electricity Requirements**	
1.5 trillion kWh		2.5 trillion kWh		3.3 trillion kWh	

availabilities were as specified in Table 8.1. The zero water price production mix for Case 8.1 consists of old gas-burning plants (20.8%), old high-sulfur coal-burning plants (62.1%), old combined-cycle plants (6.5%), and new combined-cycle plants (10.6%). This production mix was not altered over the entire range of water withdrawal prices studied. Once-through cooling was the prevalent method of waste heat removal.

Figure 8.1 indicates the price changes necessary to cause water use configuration changes within the electric power industry under conditions of low demand for electric power and minimal wastewater discharge restrictions. The figure also indicates the type of water conservation alternatives being selected by the model to reduce withdrawals. Incorporation of wet recycle

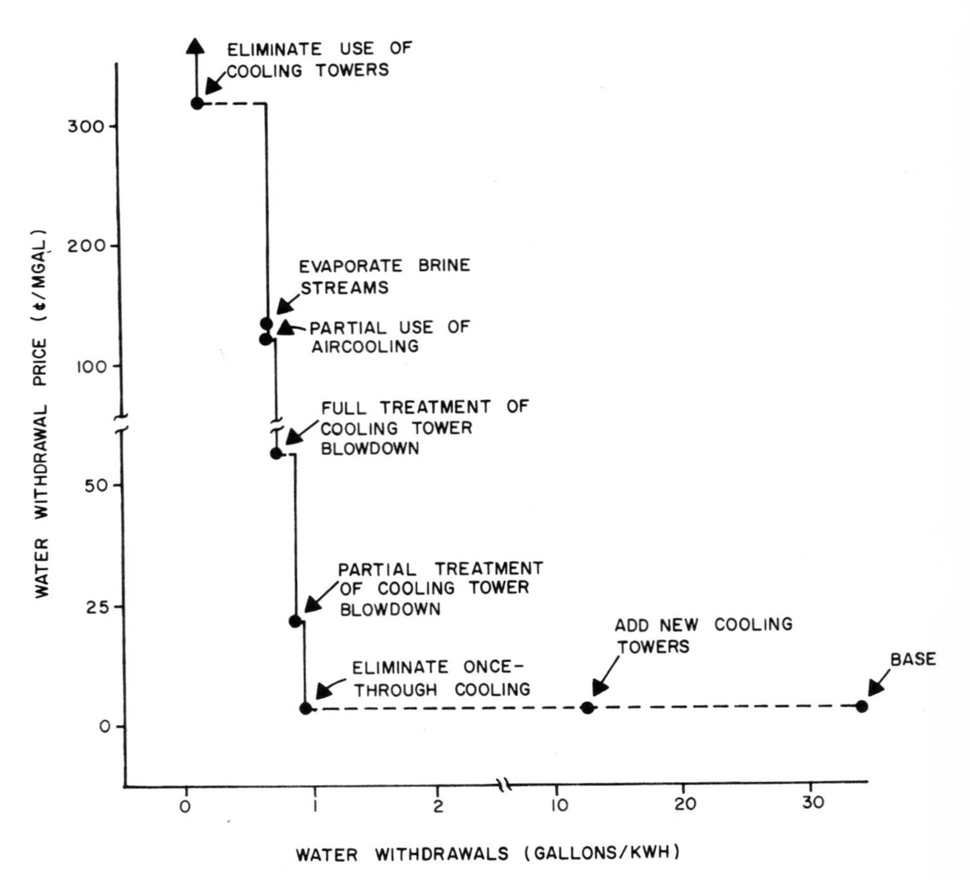

Figure 8.1. Water withdrawal price vs. water withdrawals—Case 8.1.

cooling towers causes the most dramatic reduction in withdrawals. Withdrawals are reduced from 34.2 gal./kWh to 0.88 gal./kWh, a reduction of 97%. Once this configuration change is implemented, the demand for water withdrawals is very inelastic. Remaining water conservation alternatives consist of recycling waste streams and air cooling, but the marginal cost of conservation is extremely high. Production costs increase by 3%, from 1.31¢/kWh for a zero water price to 1.35¢/kWh for the all-cooling-tower configuration. Production costs increase by another 10% to 1.49¢/kWh for the air-cooling configuration. The corresponding water withdrawal prices which induced these configurations were 1.4¢/Mgal. and $3.21/Mgal., respectively.

In case 8.2, electric power demand is raised to the medium level of 2.5 trillion kWh, the availability of natural gas is limited to approximately that needed for use in stack gas scrubbers, and environmental restrictions are tightened to impose new source standards on new plants. These environmental restrictions include use of cooling towers in new plants. Because of the different demand, fuel, and environmental assumptions, the production mix for a zero water price case changes considerably from Case 8.1. The mix is comprised of old high-sulfur coal-burning plants (53.2%), new low-sulfur coal-burning plants (27.7%), new medium-sulfur coal-burning plants (12.6%), and new combined-cycle plants (6.5%). As in Case 8.1, the mix remains unchanged over the entire range of water prices. Imposition of restrictive standards causes cooling towers to be used in a greater proportion at a zero water withdrawal price than in Case 8.1.

Figure 8.2 shows the effects of increasing water withdrawal price for specifications of medium demand and environmental restrictions. The effects differ little from Case 8.1; water withdrawal prices of the same magnitude generally cause the same process changes in the model. Assuming a zero water price, production costs were 1.49¢/kWh; costs increased by 2%, to 1.52¢/kWh for the all-cooling-tower configuration and 1.72¢/kWh for the air-cooling configuration.

In both Cases 8.1 and 8.2, heat discharges to the water are eliminated when the once-through cooling option is replaced by wet recycle cooling towers. Since primary treatment of the inlet water is the major source of water pollutants, waste discharges to the water are reduced coincident with reduced withdrawals. However, zero discharge of these pollutants (dissolved and suspended solids) is never achieved regardless of the water withdrawal prices imposed. Other studies indicate that these components can be more effectively and economically controlled by limiting the discharges directly as discussed in Chapter 6 (Calloway et al., 1974).[1] Similar results are indicated here from the analysis conducted in Case 8.2; zero discharges to water are achieved at a production cost of 1.35¢/kWh as compared to the *minimum*

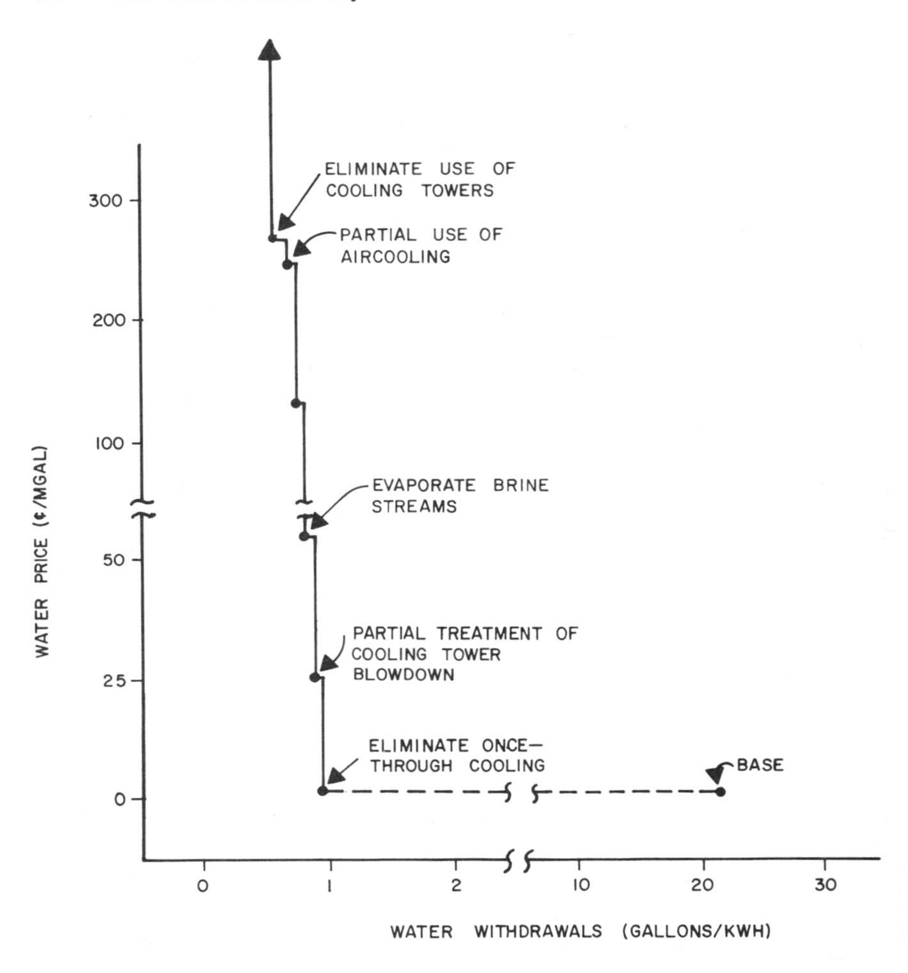

Figure 8.2. Water withdrawal price vs. water withdrawals—Case 8.2.

discharge costs of 1.49¢/kWh for Case 8.1 and 1.72¢/kWh for Case 8.2. Withdrawal prices were $3.21/Mgal. and $2.57/Mgal., respectively.

Case 8.3 is the most stringent case; electric power demand is 3.3 trillion kWh, natural gas and low-sulfur coal supplies are minimal, oil is not available, and maximum treatment is required for both air and water waste discharges. Heat discharges to the water are restricted also; thus an all-cooling-tower configuration is assumed for the initial model solution.

The production mix includes old low- (10.3%), medium- (16.0%), and high-sulfur (11.0%) coal-burning plants, new high-sulfur coal-burning plants

(1.5%), old combined-cycle plants (.3%), and coal gasification combined-cycle plants (60.9%). Power demand, fuel availability, and discharge conditions are so severe that minimal withdrawals are made even at a zero water withdrawal price. All water-saving alternatives except air cooling are being exercised for the purpose of meeting the environmental constraints. For this case, air cooling becomes feasible at a much higher withdrawal price ($6.79/Mgal.) than for Cases 8.1 and 8.2. Production costs are 2.68¢/kWh for a zero water price and 3.06¢/kWh for the air-cooling configuration.

In summary, the most cost-effective means of reducing water withdrawals in electric power generation is through use of cooling towers. Increasing water withdrawal price to force use of other water withdrawal-saving alternatives is economically unproductive. Increasing water withdrawal price to control wastewater discharges is similarly unproductive.

Derived Demand Schedules for Low-Sulfur Coal

Four different analyses were made to evaluate the effect of increasing low-sulfur coal prices on the use of low-sulfur coal in the production of electricity. Specifications for these four evaluations are summarized in Table 8.2. No limitations were placed on the availability of capital in making these four analyses. Similarly, no limitations were placed on the availability of water. One ton of low-sulfur coal was assumed to be the equivalent of 17 MMBtu's of energy.

The demand schedule estimated for Case 8.4 (not shown graphically) is simple but interesting. At the base price of 61¢/MMBtu, 11 quadrillion Btu's of low-sulfur coal were used. In comparison, total use of low-sulfur coal in 1972 was less than one quadrillion Btu's. However, use of low-sulfur coal decreased with increasing price until it became zero for prices higher than $2.23/MMBtu. Demand schedules for Cases 8.5 and 8.6 were estimated for low-sulfur coal prices ranging from 61¢/MMBtu to nearly $3.00/MMBtu. Low-sulfur coal use was 64% greater in Case 8.5 than in Case 8.4 because of less petroleum fuel availability; low-sulfur coal use was 30% greater in Case 8.6 than in Case 8.5 because of the increased restrictiveness of waste discharge standards. Specifically, decreasing the availability of clean petroleum fuels by 6.1 quadrillion Btu's increased the use of low-sulfur coal by 7.0 quadrillion Btu's.

The estimated demand schedule for low-sulfur coal for Case 8.5 is plotted in Figure 8.3. This demand curve shows that the use of low-sulfur coal in electric power generation, even with an assumed scarcity of petroleum fuels, is extremely sensitive to prices of low-sulfur coal greater than $1.80/MMBtu. Use of low-sulfur coal decreases by 54% from 16.8 to 7.4 quadrillion Btu's coincident with an 18¢ price increase from $1.80 to $1.98/MMBtu. A price of $2.85/MMBtu precludes the use of low-sulfur coal in the model.

Table 8.2. Specifications of Fuel, Waste Discharge, and Electricity Requirements for Cases 8.4, 8.5, 8.6, and 8.7 of Low-Sulfur Coal Demand Analyses

Case 8.4		Case 8.5		Case 8.6		Case 8.7	
1. Fuel Availabilities (10^{12} MBtu)		**1. Fuel Availabilities (10^{12} MBtu)**		**1. Fuel Availabilities (10^{12} MBtu)**		**1. Fuel Availabilities (10^{12} MBtu)**	
Natural gas	5.7	Natural gas	1.3	Natural gas	1.3	Natural Gas	1.3
Low-sulfur coal	—	Low-sulfur coal	—	Low-sulfur coal	—	Low-sulfur coal	—
Medium-sulfur coal	6.0	Medium-sulfur coal	6.0	Medium-sulfur coal	6.0	Medium-sulfur coal	6.0
High-sulfur coal	Unlimited	High-sulfur coal	Unlimited	High-sulfur coal	Unlimited	High-sulfur coal	Unlimited
Low-sulfur oil	.2	Low-sulfur oil	Zero	Low-sulfur oil	Zero	Low-sulfur oil	Zero
High-sulfur oil	1.5	High-sulfur oil	Zero	High-sulfur oil	Zero	High-sulfur oil	Zero
2. Water Standards		**2. Water Standards**		**2. Water Standards**		**2. Water Standards**	
Existing conditions for old plants		Existing conditions for old plants		Cooling Towers		Cooling towers	
Cooling towers for new plants		Cooling towers for new plants		Zero discharge of pollutants		Zero discharge of pollutants	
3. Air Standards		**3. Air Standards**		**3. Air Standards**		**3. Air Standards**	
Existing standards for old plants		Existing standards for old plants		Particulate control		Particulate control	
New source standards for new plants		New source standards for new plants		Stack gas scrubbers		Stack gas scrubbers	
4. Electricity Requirements		**4. Electricity Requirements**		**4. Electricity Requirements**		**4. Electricity Requirements**	
2.5 trillion kWh		2.5 trillion kWh		2.5 trillion kWh		3.3 trillion kWh	

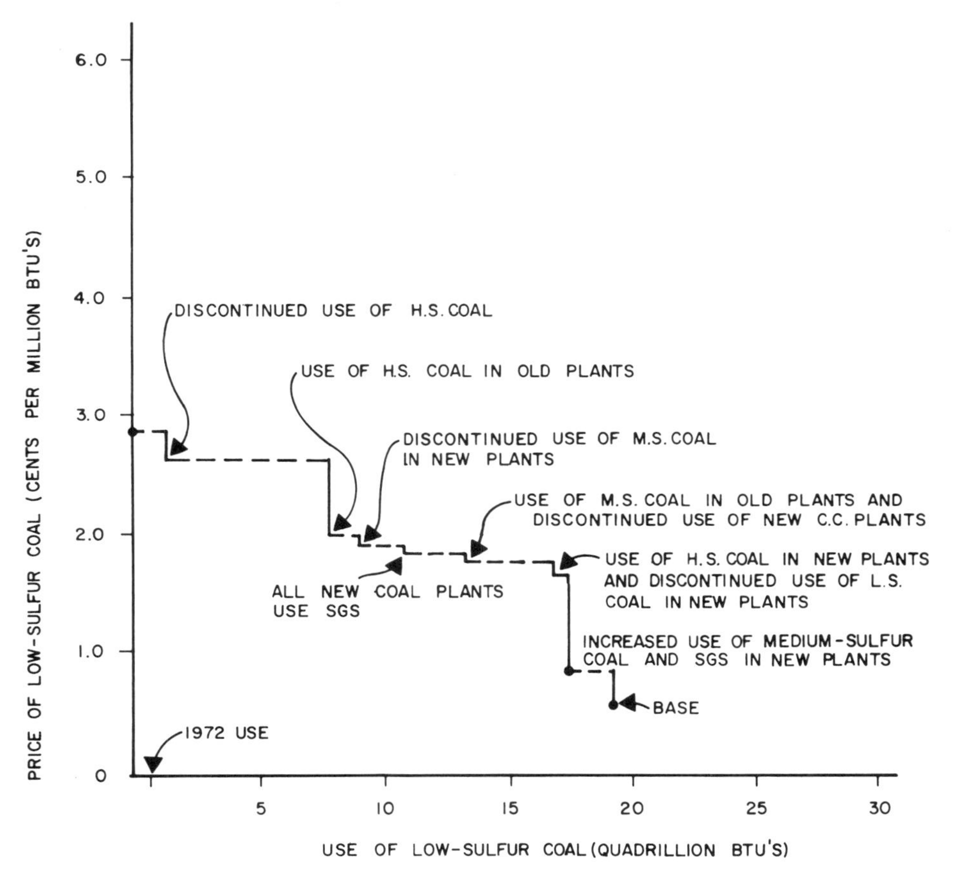

Figure 8.3. Demand schedules for low-sulfur coal for minimum petroleum fuel availabilities, moderate waste discharge restrictions, electricity requirements of 2.5 trillion kWh.

Use of low-sulfur coal differed only slightly between Cases 8.5 and 8.6 for all low-sulfur coal prices between 61¢ and $2.00/MMBtu. Significant differences in use of low-sulfur coal in these two cases existed only at the very extremes of the price range.

In Case 8.7 natural gas is limited to the minimal amount required to obtain a feasible solution of the model; no fuel oil is available. Stack gas scrubbing of sulfur dioxide emissions and precipitation of particulate discharges are required. Cooling towers are required to dissipate the heat; no pollutants may be discharged to the waterways. Electricity requirements are at the high level of 3.3 trillion kWh.

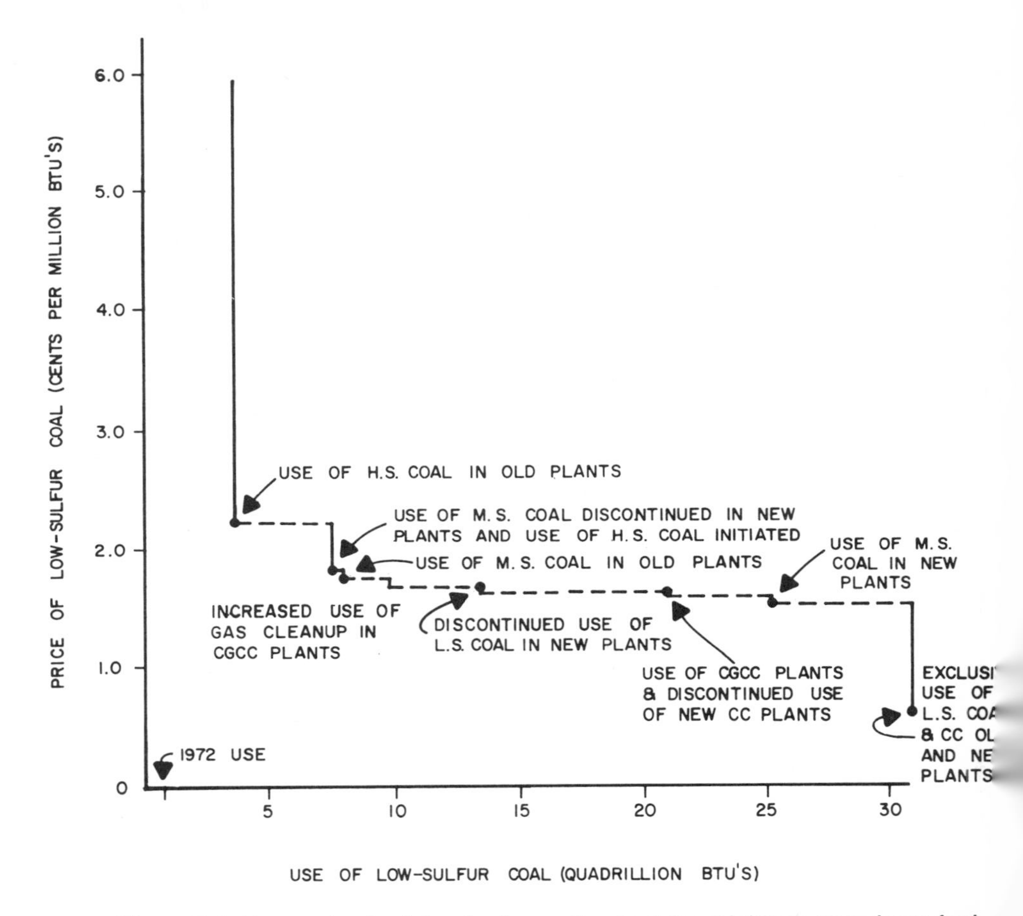

Figure 8.4. Demand schedules for low-sulfur coal for minimum petroleum fuel availabilities, stringent waste discharge restrictions, electricity requirements of 3.3 trillion kWh.

An estimate of the demand schedule for Case 8.7 is plotted in Figure 8.4. This demand curve indicates the magnitude of low-sulfur coal use, particularly at low prices of low-sulfur coal, if the most expansive combinations of demand and environmental conditions occurred for low-sulfur coal. Use of low-sulfur coal is 30.6 quadrillion Btu's at a price of 61¢/MMBtu. This level of use is 31% greater than the largest use found in Case 8.6 at the same price; it is five times greater than any level of use evaluated in the environmental and supply analyses of Chapters 6 and 7.

Use of low-sulfur coal decreases 19% with an increase in the price from 61¢ to $1.56. The model substitutes medium-sulfur coal in new plants for low-sulfur coal. Use of low-sulfur coal decreases another 17% with an increase in the price from $1.56 to $1.626/MMBtu. This is accomplished in the model by investing in coal gasification combined-cycle plants rather than investing, as before, in new combined-cycle plants. Increasing the price from $1.626 to $1.632/MMBtu's decreased the use of low-sulfur coal another 36%. Investments in new low-sulfur coal plants are discontinued. Use of low-sulfur coal at this price of $1.632 is 64% less than use at a price of 61¢.

Higher prices of low-sulfur coal result in additional changes in the generation mix at $1.76, $1.78, $1.80, and $2.11/MMBtu. Medium-sulfur coal is substituted for low-sulfur coal in old plants; use of high-sulfur coal is initiated in new plants, and use of medium-sulfur coal is discontinued in new plants; high-sulfur coal is substituted for natural gas in old combined-cycle plants at the low-sulfur coal price of $2.11/MMBtu; natural gas released from power generation use is used to operate the stack gas scrubbers.

Two important points follow from the modeling analysis of the demand for low-sulfur coal in electric power generation. First, the use of low-sulfur coal at a low price of 61¢/MMBtu is affected significantly by the availability of petroleum fuels, the restrictiveness of sulfur dioxide standards, and the level of electricity requirements. Second, the use of low-sulfur coal is extremely sensitive to low-sulfur coal prices higher than 61¢ and less than $2.00/MMBtu ($10.37 to $34/ton). Thus, according to the model, development and further growth of the low-sulfur coal resources will likely depend on the availability of petroleum fuels, the price of low-sulfur coal, and the rate at which air emission control policy is implemented and enforced.

References

1. James A. Calloway, A.K. Schwartz, Jr. and Russell G. Thompson, "Industrial Economic Model of Water Use and Waste Treatment for Ammonia," *Water Resources Research*, V. 10, No. 4, pp. 650-658, August, 1974.

9. Capital Demand

H. Peyton Young • L. Ted Moore
James A. Calloway • Russell G. Thompson

Investment capital is an essential input to electric utilities' ability to build new capacity, replace old capacity, implement technological changes, and install pollution abatement equipment to meet environmental standards. The linear-electric power model was used to analyze the effects of limited availability of capital on (1) production cost, (2) production configuration, (3) pollution abatement strategies, and (4) the marginal value of scarce fuels. An expanded version of the model was also used to derive demand curves for total utilities' investment capital under different assumptions about electric power demand, fuel availabilities, and environmental policies; these demand curves show how the marginal value of capital to utilities increases as capital becomes scarcer.

To evaluate the total capital requirements of the electric power industry to 1985, the electric power model was expanded to include all electric power equipment requiring major investments; this expansion includes nuclear and hydroelectric generation capacity in addition to all previously modeled fossil-fueled capacity. Peaking units were included to provide an assumed 5% of 1985 electrical generation. Peaking units include internal combustion engines and combustion turbines. Hydroelectric generation was assumed to provide 0.3×10^{12} kWh in 1985. Nuclear generation was included but was allowed to provide no more electricity than current building plans indicate is possible (*Electrical World*, March 15, 1975).[1]

For the purpose of adding further investment flexibility to the model, a Wellman-Lord sulfur dioxide scrubber (*Electrical World*, November 1, 1972)[2] was modeled as an alternative to the wet limestone process. In addition, the capital cost of converting oil- and gas-burning plants to coal-burning plants, which was not included in the previous analysis, has been included in the model. This cost was assumed to be $90/kW. Transmission and distribution capital costs are treated as a percentage of total new generating plant capital expenditures. This percentage was estimated from historical data; the model estimates transmission and distribution capital requirements at 70% of new generating plant costs.

The addition of specific units for peaking generation causes the model's requirements for low-sulfur fuels to be higher; to add flexibility in fuel use to the model, peaking units and stack gas scrubbers were allowed to burn either natural gas or low-sulfur oil.

Capital Limitations

Based on projections of total investment capital available to business in the period 1975-1985 and the capital share that the utilities have successfully bid for in the past, capital available for utilities was estimated to total between 163 and 261 billion dollars (1973 dollars). The latter figure assumes a high rate of GNP growth, the former a low rate of growth (see Chapter 4). These figures are also predicated on the assumption that the electric utility sector will not be able to increase its share of the capital market primarily because of its generally weak financial condition relative to other bidders for capital. If regulatory agencies were to allow significantly higher rates of return, or if long-term interest rates were to decline substantially, then these bounds may be too low.

Capital Costs and Substitution Possibilities

The model illustrates how the electric power industry can respond to increasingly severe capital shortages by *substituting* other inputs for capital. The patterns of substitution depend heavily on the relative prices of the various types of capital inputs and their relation to the prices of all other inputs, as well as on the structure of the model itself. Table 9.1 gives the cost/kW of installed capacity of the major capital components used in the model.

The pattern of substitution for capital as capital supplies are reduced may be described generally as follows. New construction will be reduced, and older, less efficient plants will be used instead of new ones. This substitution is limited, however, by the production capacity of the older plants.

Strict environmental standards will require new capital spending for pollution control equipment. Hence, as capital availability is reduced, other means

Table 9.1. Major Capital Cost Components

Unit	Cost (1973 dollars per kW)
Coal-fired fossil plant	230
Oil-fired fossil plant	228
Gas-fired fossil plant	200
Combined-cycle plant	180
Coal gasification combined-cycle plant	254
Cooling tower	7
Scrubber (new plant)	40
Scrubber (retrofitted to existing plant)	70
Particulate treatment unit	5.50

of meeting environmental requirements will have to be used. However, substitution possibilities of this type are limited. To meet SO_2 effluent restrictions, some utilities could switch to greater use of low-sulfur fuels (low-sulfur oil and natural gas) and install fewer scrubbers, thus reducing capital expenses, but operating expenses would increase since low-sulfur fuels would be more expensive, and this substitution possibility is limited in scope because of the limited availabilities of clean fuels.

If environmental controls are exercised by taxing effluents instead of placing restrictions on them, there is greater freedom to respond to capital shortages. Namely, as capital becomes scarce, more utilities would simply pay the taxes instead of investing in capital-intensive clean-up equipment.

In general, as capital becomes more scarce, the marginal value of capital increases; that is, the utilities will be willing to pay higher interest rates to obtain capital. Scarcity of capital leads to a demand curve for marginal capital as a function of capital availability. This demand curve changes with assumptions concerning resource availabilities and power generation requirements. A lower bound exists where capital is too scarce to meet the various production constraints; below this "cut off" value, the model becomes infeasible.

Three Cases for Capital Analysis

The effects of capital limitations in the range of $261 billion and lower were investigated for three different sets of assumptions concerning fuel availabilities and prices, environmental policies, and electric power demand. These three cases are summarized in Table 9.2.

Case 9.1 is the least restrictive of the environmental cases analyzed; however, electricity requirements in Case 9.1 are 4.12 trillion kWh (not including in-plant use), with nuclear generation representing at most .52 trillion kWh of this total. This case may be called a "minimum environment, low nuclear option" case.

Case 9.2 is similar to Case 9.1 except that maximum environmental standards are assumed for both air and water discharges. Electric power demand is assumed to be 3.36 trillion kWh and nuclear generation is allowed to range as high as 1.51 trillion kWh. This case may be characterized as the "maximum environment, high nuclear option" case.

Case 9.3 assumes the same fuel availabilities and prices as in the above cases, but discharges of SO_2 and particulates are priced at $0.50/lb. and $2.50/lb., respectively; water withdrawal is priced at $0.03/Mgal. A water withdrawal price of $0.03/Mgal. insures use of wet cooling towers; otherwise, waste discharges to the water are unrestricted. Total electric power demand is 4.1 trillion kWh, of which (at most) 1.3 trillion kWh is allowed to be nuclear. This could be called the "effluent tax, medium nuclear option" case.

Table 9.2. Three Cases for Capital Demand Analysis

Electric Power Demand (10^{12} kWh)	Case 9.1 Minimum Environment, Low Nuclear	Case 9.2 Maximum Environment, High Nuclear	Case 9.3 Effluent Tax, Medium Nuclear
Fossil	3.30	1.55	2.50
Nuclear	.52	1.51	1.30
Hydro	.30	.30	.30
Total	4.12	3.36	4.10
Costs			
Natural gas (¢/MMBtu)	64	64	64
Low-sulfur oil (¢/MMBtu)	139	139	139
High-sulfur oil (¢/MMBtu)	119	119	119
Low-sulfur coal (¢/MMBtu)	61	61	61
Medium-sulfur coal (¢/MMBtu)	50	50	50
High-sulfur coal (¢/MMBtu)	44	44	44
Water (¢/1000 gal.)	0	0	3
Availabilities (10^{12} MBtu)			
Natural gas	5.7	5.7	Unlimited
Low-sulfur oil	.2	.2	Unlimited
High-sulfur oil	1.5	1.5	Unlimited
Low-sulfur coal	6.16	6.16	Unlimited
Medium-sulfur coal	6.00	6.00	Unlimited
High-sulfur coal	Unlimited	Unlimited	Unlimited
Water Standards			
	None	Zero discharge of dissolved and suspended solids	Water price 3¢/Mgal.
Air Standards			
	None	Zero discharge of SO_2 particulates	SO_2 tax of \$.50/lb. Particulate tax of \$2.50/lb.

Effects of Capital Limitations

The model shows that, generally, as capital availability is reduced, the marginal value of capital (the amount utilities would be willing to pay for an extra dollar of capital) increases dramatically. This marginal value, measured in 1973 dollars, may be loosely interpreted as the real additional interest that utilities would be willing to pay to borrow an additional dollar for new investment. When added to the interest rate on long-term debt that the utility industry already pays (about 7.1% in 1974, see Table 9.12), this marginal value gives a measure of the utilities' demand for investment capital. The problem of depreciation on fixed investment, as well as the utilities' ability or inability to expand its debt, is not taken into account in the model.

The demand curves obtained in this way for Cases 9.1, 9.2, and 9.3 are illustrated in Figures 9.1 and 9.2. These demand curves represent average demand curves for the utility industry based on 1973-1974 conditions; they do not reflect the abilities of particular utilities to borrow in the capital markets. Thus, although the curves contain much relevant information, they should be interpreted with some caution.

First, the marginal value of capital increases at an increasing rate as capital availability is curtailed. In Figure 9.1, as capital is reduced from $172 billion to $168.7 billion in Case 9.1, the marginal value increases incrementally from zero to 3.3¢/dollar (i.e., 3.3%). A further slight decrease in capital availability from $168.7 to $168 billion results in a striking increase in the marginal value of capital from 3.3¢ to 37¢/dollar. Finally, if available capital is less than $165.3 billion, then the requirements of Case 9.1 cannot be met by any process adjustment in the model. A similar pattern of increasing marginal values is seen in Case 9.2 (Figure 9.1) and Case 9.3 (Figure 9.2). The minimum capital requirements in Cases 9.1 and 9.2 are both greater than the lower-bound forecast of $163 billion discussed in Chapter 4.

Capital requirements in Case 9.3 are $142 billion at a maximum, which is considerably less than the lower-bound forecast of capital availability. Because the model has the option to pay the effluent taxes and purchase clean fuels rather than make the investments in waste control equipment, the minimum capital requirements in Case 9.3 are only $129.6 billion. The public in this case does not have to abstain nearly as much from consumption to make the savings available for capital investment; however, the public must pay the higher incidence costs of the pollution and must import greater quantities of crude oil.

A second important feature of these demand curves is that they show process shifts which occur as capital constraints become tighter. In other words, the model solutions determine the substitutions that are being made for capital as capital becomes more scarce. These substitutions are of two general types: shifts in the types of plants used to produce power (implying a shift in the types of fuel used) and shifts in the types of processes used to meet environmental standards (this may also affect the "mix" of fuel types used and hence their marginal values).

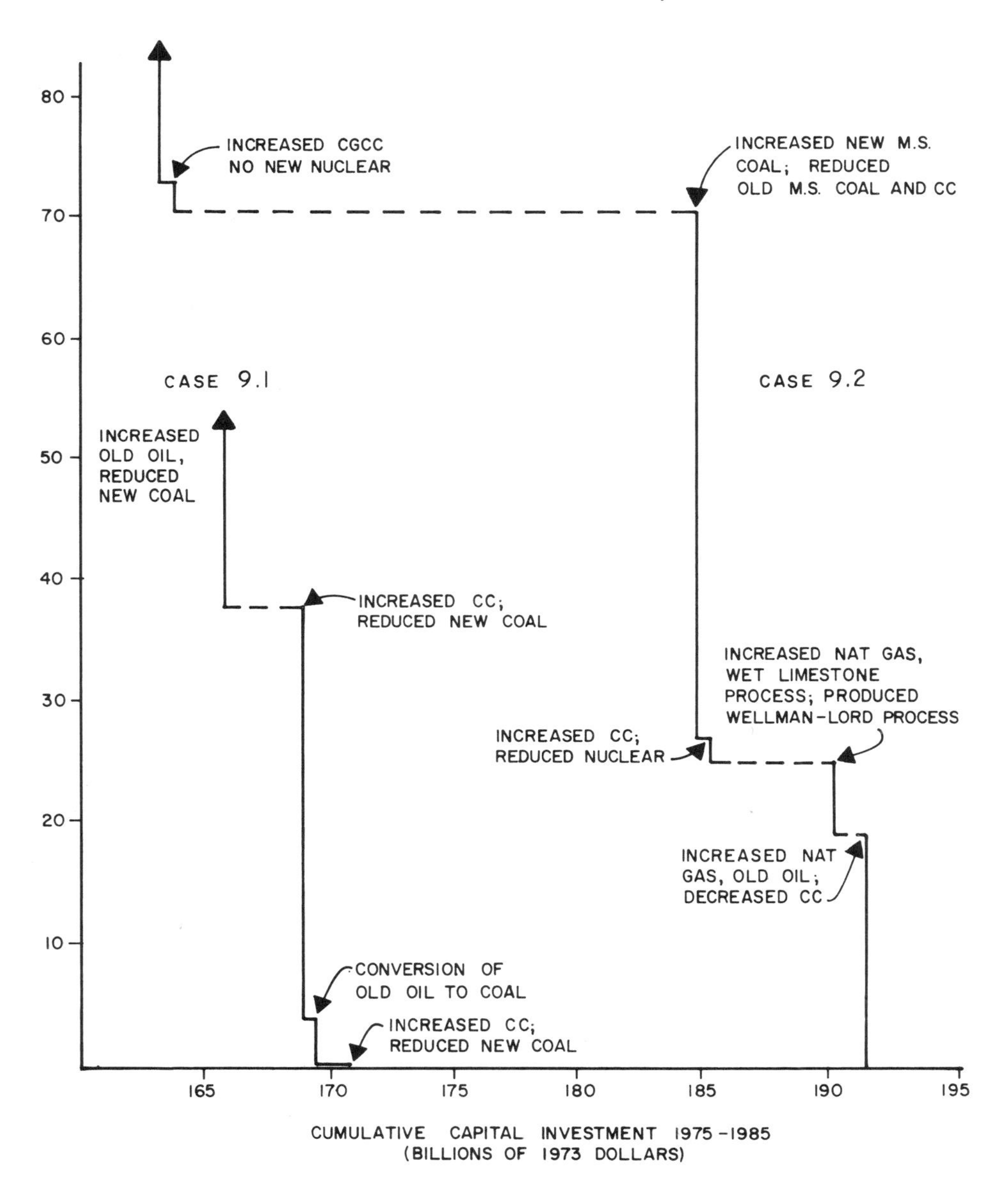

Figure 9.1. The relationship between the marginal value of capital and cumulative capital investment for Cases 9.1 and 9.2.

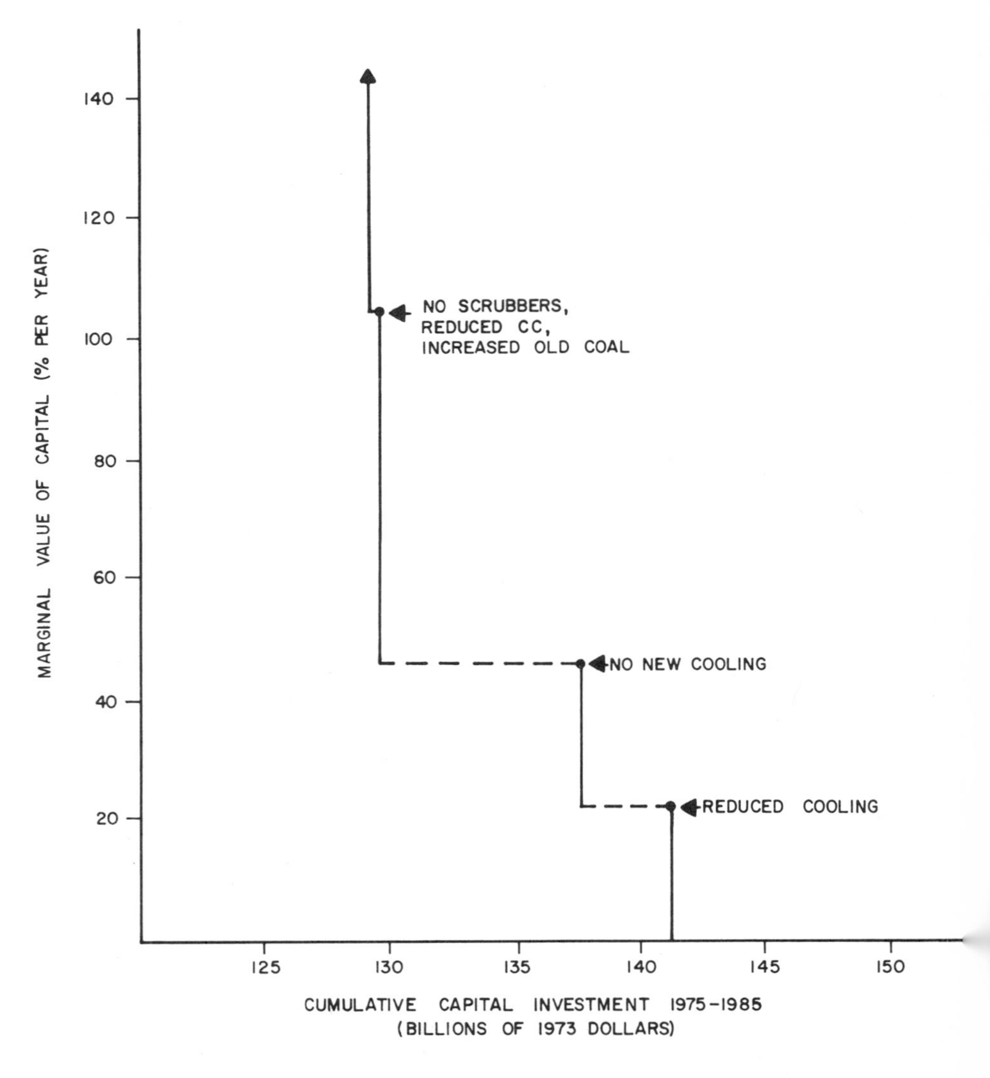

Figure 9.2. Estimated demand for capital—Case 9.3.

Table 9.3 shows the changes in the production mix caused by shifts in capital under Case 9.1. Table 9.4 shows the shifts in the marginal values of the scarce fuels resulting from the same variations in capital. Referring to these tables and to Figure 9.1, one can see that in the base case where capital availability is $175 billion, high-sulfur coal is used exclusively in new plants

Table 9.3. Capital-Induced Shifts in Production Mix—Case 9.1

	Production (in-plant use and sale in trillion kWh)				
	4.2	4.2	4.2	4.2	4.2
	Capital ($ in billions)				
	175.0	171.7	168.7	168.0	165.3
Type of Plants	Percent of Generation Mix				
New Plants					
High-sulfur coal	52.3%	46.8%	45.1%	44.4%	44.4%
Combined-cycle	—	5.4	5.4	6.1	6.1
Peaking	3.3	3.3	3.3	3.3	3.3
Old Plants					
High-sulfur coal	30.4	30.4	31.8	31.6	28.5
High-sulfur oil	—	—	—	0.3	3.4
Combined-cycle	2.3	2.3	2.3	2.3	2.3
Nuclear	2.5	2.5	2.5	2.5	2.5
Hydro	7.1	7.1	7.1	7.1	7.1
Peaking	1.9	1.9	1.9	1.9	1.9

Table 9.4. Capital-Induced Shifts in the Marginal Values of Scarce Fuels—Case 9.1

($/MMBtu)*

	Capital				
Fuel Type	175.0	171.7	168.7	168.0	165.3
Natural gas	0	.014	.064	.76	1.32
High-sulfur oil	0	0	0	0	1.25
Low-sulfur oil	0	0	0	.01	1.31

*Marginal values of the coal types are zero in all cases.

(except for peaking), and is the predominant fuel in old plants as well. This implies some fuel-switching by old plants to the use of coal because high-sulfur coal is the cheapest fuel to burn. However, as capital is decreased to $171.7 billion, the model shifts to some use of combined-cycle in new plants. This saves capital because new combined-cycle plants are cheaper to build per kW of capacity than are new fossil-fuel plants (see Table 9.1). However, the amount of new combined-cycle capacity is sharply limited by the supply of natural gas. As capital becomes still more scarce, fewer old oil-burning plants are converted to coal-burning plants; hence the use of high-sulfur (old) oil plants increases somewhat in the model results. As shown in Table 9.4, the marginal value of natural gas rises rapidly as available capital declines. The model would therefore elect to build more combined-cycle plants if natural gas were available to operate them.

In Case 9.1, there are no shifts in investment waste control technology since these investments are always zero when there are no environemtal constraints. Note that the overall system is not very flexible with regard to capital variations: at $175 billion the model does not even use all the capital, while at $165 billion (a decrease of less than 6%) the model is infeasible. On the other hand, it is interesting to note that the lower feasible bound of $165.3 billion corresponds with the estimated lower bound on capital in Chapter 4.

Case 9.2 deals with the effects of capital limitations under stringent environmental standards. Figure 9.2 shows the effects of increased capital scarcity on unit production costs. Table 9.5 shows how operating costs increase with decreased capital availability; a decrease in available capital of $9.1 billion over the period 1975-1985 (an average of $.83 billion/yr. leads to an increase in 1985 operating costs of $2 billion. Table 9.6 shows the production mix changes as capital is decreased; Table 9.7 shows the corresponding changes in waste control processes. In the base case, new generation is essentially limited to combined-cycle and nuclear plants. As capital becomes scarce, the first response, at $191.2 billion, is to reduce construction of combined-cycle plants and to increase the utilization of old plants, particularly natural gas. At $190.1 billion, a change occurs in the SO_2 removal technology—the wet limestone process is substituted for the Wellman-Lord process because the latter requires more capital investment in process steam boilers.

At $185.3 billion, new combined-cycle plants are brought in again instead of new nuclear plants, but at $184.9 billion another solution is found—namely, to build more coal plants (with scrubbers) and avoid the extra cost of using old plants burning coal. The capital expended in switching old plants

Table 9.5. Capital Availability and Operating Costs for Case 9.2

($ billion)

Capital Availability	Operating Costs
194.0	49.3
191.2	49.8
190.1	50.0
185.3	51.2
184.9	51.3
163.5	66.7

from gas to coal is saved along with the extra costs of retrofitting scrubbers to old plants. As a result, available gas is burned in old gas-burning plants instead of new combined-cycle plants. Finally, the model chooses to invest CGCC instead of nuclear plants; this also reduces the need for cooling towers and reduces capital needs by $20 billion, or more than 10%. The steady upward trend of the marginal values of fuels as capital becomes more scarce is illustrated in Table 9.8

Table 9.6. Capital-Induced Shifts in Production Mix for Case 9.2

	Production (in-plant use and sale in trillion kWh)					
	3.55	3.55	3.55	3.55	3.55	3.62
	Capital ($ billion)					
	194.0	191.2	190.1	185.3	184.9	163.5
Type of Plant	**Percent of Generation**					
New Plants						
Medium-sulfur coal	—	—	—	—	5.5%	4.4%
Combined-cycle	4.8%	3.0%	3.2%	7.5%	0.8	—
CGCC	—	—	—	—	—	39.1
Nuclear	39.5	39.5	39.5	35.2	36.3	—
Peaking	3.0	3.0	3.0	3.0	3.0	3.0
Old Plants						
Gas or oil	1.8	2.5	0.9	3.9	3.9	3.8
Natural gas	0.7	1.9	3.5	0.6	5.1	4.2
Medium-sulfur coal	16.4	16.4	16.4	16.4	11.9	12.5
Low-sulfur coal	16.9	17.1	17.1	17.1	16.9	16.6
Combined-cycle	2.8	2.8	2.8	2.8	2.8	2.7
Nuclear	3.0	3.0	3.0	3.0	3.0	3.0
Hydro	8.5	8.5	8.5	8.5	8.5	8.3
Peaking	2.3	2.3	2.3	2.3	2.3	2.2

Table 9.7. Capital-Induced Shifts in Environmental Control Investment for Case 9.2

	Capital ($ billions)					
	194.0	**191.2**	**190.1**	**185.3**	**184.9**	**163.5**
Wellman-Lord process	17.3	18.0	—	—	—	—
Wet limestone process	—	—	16.5	19.1	16.9	17.3
Particulate removal	—	—	—	—	—	—
Cooling towers	4.3	4.3	4.2	4.1	4.2	2.7
Total Treatment	21.6	22.3	20.7	23.2	21.1	20.0

Table 9.8. Capital-Induced Shifts in the Marginal Values of Scarce Fuels—Case 9.2

	Capital ($ billion)					
	194.0	191.2	190.1	185.3	184.9	163.5
Fuel Type	Marginal Value ($/MMBtu)					
Natural gas	1.44	1.98	2.13	2.27	5.47	5.74
Medium-sulfur coal	.52	.18	.08	.13	1.40	1.44
Low-sulfur coal	.86	.86	.85	.93	3.00	3.07
High-sulfur oil	0	0	0	0	0	0
Low-sulfur oil	.69	1.23	1.38	1.52	4.72	5.00

Table 9.9. Capital-Induced Shifts in Production Mix for Case 9.3

	Total Production (trillion kWh)			
	4.2	4.2	4.2	4.2
	Capital ($ billion)			
	142.5	141.1	137.6	129.6
Type of Plants	Percent of Generation			
New Plants				
Combined-cycle	52.4%	52.4%	51.7%	50.1%
Peaking	3.3	3.3	3.3	3.3
Old Plants				
Natural gas	7.6	7.6	7.6	7.6
Low-sulfur coal	15.8	15.8	15.8	19.3
Low-sulfur oil	7.0	7.0	7.0	5.4
Combined-cycle	2.4	2.4	2.4	2.4
Nuclear	2.5	2.5	2.5	2.5
Hydro	7.1	7.1	7.1	7.1
Peaking	1.9	1.9	1.9	1.9

Table 9.10. Capital-Induced Shifts in Environmental Control Investment for Case 9.3

	Capital ($ billion)			
	142.5	141.1	137.6	129.6
Wellman-Lord process	4.2	4.2	4.2	0
Wet Limestone process	0	0	0	0
Particulate removal	0	0	0	0
Cooling tower	3.3	1.9	0	0
Total investment	7.5	6.1	4.2	0

Case 9.3 illustrates the situation in which waste discharges are taxed instead of restricted absolutely. Solution to the model without capital constraints requires use of combined-cycle generation in new plants. SO_2 recovery (the Wellman-Lord process), particulate removal, and cooling towers are all used to avoid paying discharge taxes. As less capital is made available, the model adjusts by using old coal plants instead of building as many new combined-cycle plants and also by not building scrubbers or cooling towers. However, particulate removal continues in the old coal plants. Tables 9.9 and 9.10 summarize these shifts in production and waste management processes; these results are shown in Figure 9.3. Table 9.11 shows how

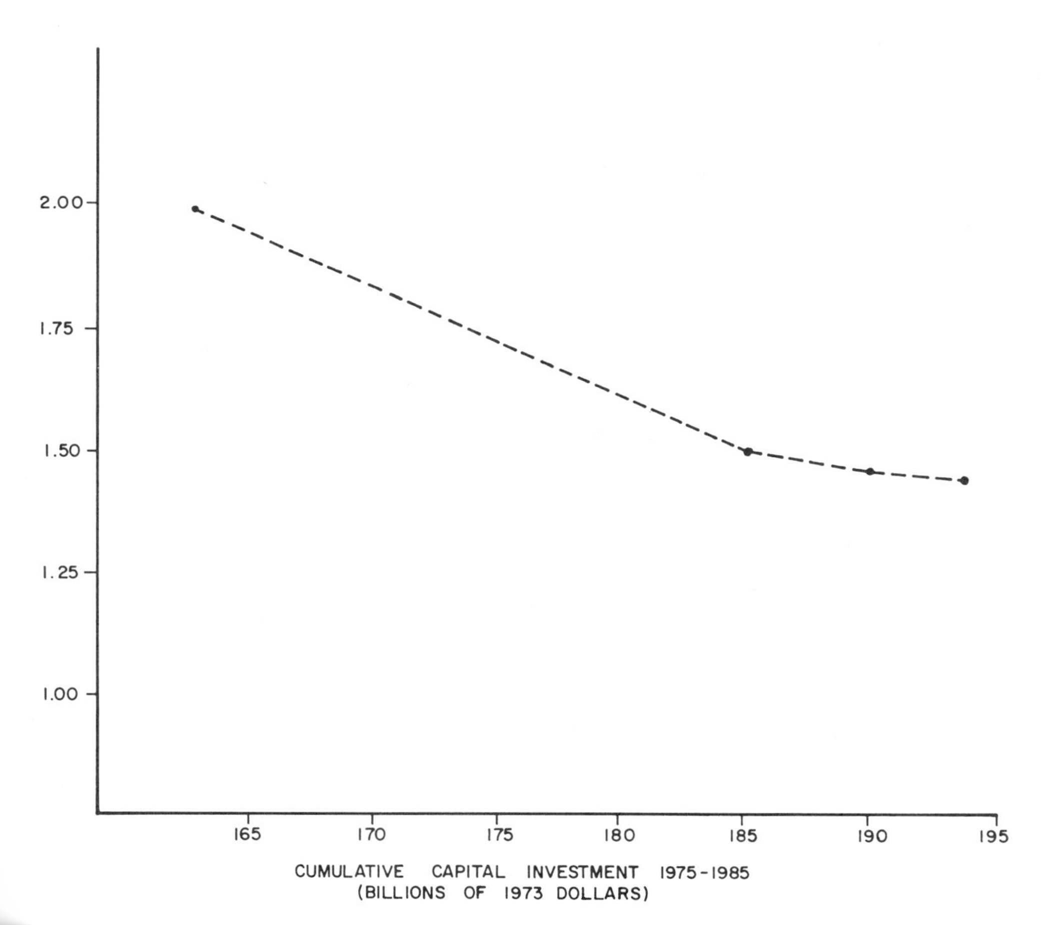

Figure 9.3. The relationship between unit costs and the availability of capital for investment—Case 9.2.

Table 9.11. Capital Costs, Operating Costs, and Pollutant Emissions for Case 9.3

Capital costs ($ billion)	142.5	141.1	137.6	129.6
Operating costs ($ billion) ..	64.3	64.9	65.7	72.9
SO_2 (lbs.)	1.31×10^{10}	1.31×10^{10}	1.31×10^{10}	1.53×10^{10}
Particulates (lbs.)	4.32×10^{9}	4.32×10^{9}	4.31×10^{9}	5.39×10^{9}
Heat to water (MMBtu)	0	5.82×10^{8}	1.34×10^{9}	1.34×10^{9}

operating costs increase and pollution emissions increase with decreased capital availability. For a decrease in available capital of $12.9 billion (an average of $1.16 billion/yr.), the 1985 operating costs increase by $8.6 billion. Clearly, the electric power industry needs considerably less capital when effluent taxes rather than effluent standards are imposed, but the consumer has to pay higher prices for electricity and tolerate the environmental degradation resulting from the utilities' inability to afford pollution control equipment.

Strict Environmental Standard Costs

Some aspects of capital requirements in Case 9.2 can be compared to corresponding capital requirements reported from recent projective studies of the Environmental Protection Agency (1974)[3] and the National Commission on Water Quality (1975).[4] These studies project the effects of enforcing strict environmental standards by 1983. The EPA study states that $12 billion would be needed to meet 1983 BAT water standards (3.3% of base capital requirements). The NCWQ report indicates that $4.56 to $9.12 billion of capital (1.9% to 3.8% of $240 billion in 1973 dollars) will be required for water pollution control by the electric utilities. Air pollution control would need 8.3% of the $240 billion over the 1973-1983 period.

The University of Houston study indicates the proportions of capital for pollution control in Case 2 are 2.3% for water clean-up and 8.9% for air clean-up. In other words, with stringent environmental standards 2 to 3% of total capital requirements for utilities will be spent for water pollution control equipment and 8 to 9% of capital will be spent for air pollution control equipment.

Variations in Fixed Charge Rates

The cost of capital, called the "fixed charge rate," enters the model explicitly as a fixed percentage each year. The fixed charge rate used in obtain ing the foregoing results is 14% of total capital spending. In actuality, the fixed charge rate is a combination of depreciation charges, return on equity, interest on long-term debt, and the tax rate. Historically, all these, except the tax rate, have varied considerably while the total fixed charge rate has remained fairly constant at 14-15%. The tax rate is virtually constant at about 50%. Table 9.12 shows the historical series for the fixed charge rate and its variable components for the period 1967-1974 (based on data from *Electrical World*, March 15, 1971 and March 15, 1975).[5,1] It is assumed that the tax rate is approximately 50% of income after deduction of depreciation, interest, and amortization; the approximate formula for computing the fixed charge rate, f, is

$$f = \frac{2\epsilon E + iD}{E + D} + d$$

where E = total equity
D = total debt
ϵ = rate of return on equity
i = interest rate on debt
d = depreciation rate

Table 9.12 shows that the return on equity has been slowly deteriorating while the interest rate has been steadily rising. If capital becomes increasingly scarce in the coming decade, interest rates will rise. Return on equity will also have to increase if utilities are to attract sufficient capital (which depends on rate increases granted by regulatory agencies).

Since the fixed charge rate may vary considerably over the period being studied (1975-1985), model solutions were computed using a range of values for the fixed charge rate to determine its impact on the type of answers obtained. In general, these variations in f were found to have little effect on the solution (except to change all capital-related costs by the same factor).

Table 9.12. Fixed Charge Rates and Components 1967-1974

Year	(f) Fixed Charge (%)	(d) Depreciation (%)	(i) Interest and Amortization (%)	(r) Return on Equity (%)
1974	15.1	2.8	7.1	9.4
1973	14.8	2.7	6.1	9.9
1972	14.7	2.7	5.9	9.7
1971	14.5	2.8	5.6	9.4
1970	14.7	2.9	5.4	9.8
1969	15.6	3.0	4.9	10.3
1968	15.8	3.0	4.4	10.4
1967	15.7	3.0	4.1	10.8

One example will suffice to show the type of shift observed. Under the assumptions of Scenario 1, with capital at $175 billion (essentially unconstrained), the only significant variation occurred when f was increased from 14% to 15%. The solutions for f = 14% and f = 15% are shown in Table 9.13.

The change induced by the increased fixed-charge rate is the shift to a less capital-intensive production method, namely, combined-cycle instead of high-sulfur coal. This change is limited, however, by the supply of natural gas.

Table 9.13. Model Solutions for Fixed Charge Rates of 14 and 15 Percent

	f = 14	f = 15
New Plants		
High-sulfur coal	2.20	1.90
Combined-cycle	—	.23
Peaking	.13	.14
Old Plants		
High-sulfur coal	1.29	1.34
Combined-cycle	.10	.10
Hydro	.30	.30
Nuclear	.11	.11
Peaking	.08	.08
Total (trillion kWh)	4.2	4.2

Summary

In summary, the capital demand analysis for the electric power industry shows that the industry has substantial flexibility to respond to limited capital availability by process substitutions. As capital becomes more scarce, the marginal value of capital rises rapidly because processes with high operating costs and low capital costs (e.g., combined-cycle plants) are substituted for processes with low operating costs and high capital costs (e.g., high-sulfur coal plants with air emission control equipment). Capital availability may significantly affect the structure of the electric power industry in the next decade. The structure of the industry may also be affected by factors included in the fixed charge rate—depreciation, interest, return on equity, and taxes. Availability of capital could inhibit the ability of the electric power industry to supply the electricity requirements of the nation if the lowest forecast of availability materializes and if present methods of implementing environmental policy are continued. Use of effluent taxes instead of effluent standards could significantly alleviate the need for capital in the electric power industry; however, the consumer must tolerate higher electricity costs and increased pollution incidence.

References

1. 1975 Annual Statistical Report, *Electrical World,* Vol. 183, Number 6, March 15, 1975.
2. "SO_2-Abatement System Builds on Success," *Electrical World*, November 1, 1972, pp. 70-72.
3. Environmental Protection Agency, *Development Document for Proposed Effluent Limitations Guidelines and New Source Performance Standards for the Steam Electric Power Generating Point Source Category*, Washington, D.C., March 1974, p. 464.
4. Governor Raymond P. Shafer, Chairman, *Report of the Domestic Council Task Force on Water Quality*, October 6, 1975, pp. A-129-A-134.
5. 1971 Annual Statistical Report, *Electrical World*, Vol. 175, Number 6, March 15, 1971, p. 39.

Appendix

Mathematical Description of the Electric Power Model

The use of linear programming to estimate (1) demand functions for the use of water, fuel and capital and (2) the cost of treating air and water pollutants is a straightforward extension of the classical application of linear programming to the firm. The electric power model describes the material flows and production processes necessary to provide the nation's supply of electric power. In general, this linear economic model shows how cost-conscious managers will respond to policy decisions limiting air and water discharges and resource availability.

Schematically, if $X_1, X_2, \ldots, X_N$ represent the levels of N production process activities, and there are a total of M types of "material flow" identified, then each activity in the model will be associated with a column vector of coefficients

$$a = \begin{bmatrix} a_{1i} \\ a_{2i} \\ \cdot \\ \cdot \\ \cdot \\ a_{Mi} \end{bmatrix}$$

such that $a_{ji} X_i$ represents the amount of good j which is input to (if negative in sign) or output from (if positive) activity i being run at level X_i. If A is the matrix of activity coefficients and X is the vector of activities, then the problem to be solved can be stated as:

$$\begin{aligned} \text{Minimize } & C^T X \\ \text{subject to } & A_E X = B_E, \\ & A_L X \leq B_L, \\ & A_G X \geq B_G, \\ & X \geq 0 \end{aligned} \qquad (1)$$

where

- E refers to the set of material balance equations
- L refers to the set of "less than" constraints
- G refers to the set of "greater than" constraints and,
- B is a vector of right hand sides.

The formulation stated in equation (1) is applicable to the electric power model. Primary model activities can be divided into four categories: exogenous supplies ($X_1 \ldots X_S$); power production alternatives ($P_1 \ldots P_J$); treatment of residuals ($T_1 \ldots T_L$) and residuals discharges ($D_1 \ldots D_V$).

Exogenous supplies (S) include fuels, water, and capital. Production alternatives (J) include existing (old) power plants and new plants. Treatment of residuals (L) includes possible use of cooling towers, stack gas scrubbers, particulate precipitators, and water recycle equipment. Residuals not eliminated by treatment are discharged to the environment.

Table 1 describes the individual activity categories in terms of their inputs and outputs. These are categorized row-wise as exogenous resource constraints, electricity demand, input resource balances, residual balances, residual discharge constraints, and finally, the objective function.

Table 1. Mathematical Description of Linear Electric Power Model

Columns 's	Exogenous Supplies $X_1 \ldots X_S$	Production Alternatives $P_1 \ldots P_J$	Treatment of Residuals $T_1 \ldots T_L$	Residuals Discharge $D_1 \ldots D_V$	Right Hand Side
genous esources	Y_1 $\ddots$ Y_S				$\leq$ Q
:ricity :mand		$e_1 \ldots e_J$			$\geq$ E
urce aterial lances	t_1 $\ddots$ t_S	$-q_{11} \cdots -q_{1J}$ $-q_{ij}$ $-q_{S1} \cdots -q_{SJ}$			= 0
ary sidual ances		$r11 \ldots r_{LJ}$ r_{ij} $r_{V1} \cdots r_{VJ}$	$-a_{11} \cdots -a_{1L}$ $-a_{ij}$ $-a_{V1} \cdots -a_{VL}$	$-d_{11} \cdots -d_{1V}$ $-d_{ij}$ $-d_{1V} \cdots -d_{VV}$	= 0
ual charge straints				1 $\ddots$ 1	$\leq$ R
ive ction	Prices of Exogenous Supplies	Costs of Production	Costs of Treatment	Effluent Charges	

Row Definitions

EAIRHT	CONSTRAINT ON STACK HEAT	MBTU
EADNAOH	50% CAUSTIC-ANHYDROUS BASIS	LB
EALUM	ALUMINUM SULFATE HYDRATE	LB
EASHWEV	WATER FROM ASH POND	GAL
EBHISC	ELECTRIC PLANT USE OF HIGH-SULFUR COAL	MBTU
EBHISO	ELECTRIC PLANT USE OF HIGH-SULFUR OIL	MBTU
EBLOSC	ELECTRIC USE OF LOW-SULFUR COAL	MBTU
EBLOSO	ELECTRIC PLANT USE OF LO-SULFUR OIL	MBTU
EBMEDSC	ELECTRIC USE OF MEDIUM-SULFUR COAL	MBTU
EBOILBLO	BOILER BLOWDOWN FOR DISPOSAL	GAL
ECLARW	BALANCE FOR CLARIFIED WATER	GAL X 10
ECLGASH	CLEAN GAS FROM DESULFURIZER - HIGH SULFUR	MMBTU
ECLGASL	CLEAN GAS FROM DESULFURIZER - LOW SULFUR	MMBTU
ECLGASM	CLEAN GAS FROM DESULFURIZER - MEDIUM SULFUR	MMBTU
ECLPART	CONSTRAINT ON UNTREATED PARTICULATES	LB X E-
ECLSO2	CONSTRAINT ON UNTREATED SO2	LB X 10
ECLWATER	DECANTED WATER TO PRECIPITATORS	GAL X 1
ECOALBRN	COAL USED FOR ELECTRICITY	MBTU
ECONS	ACCOUNTING ROW FOR WATER CONSUMPTION	GAL
ECRGASH	CRUDE GAS FROM DESULFURIZER - HIGH SULFUR	MMBTU
ECRGASL	CRUDE GAS FROM DESULFURIZER - LOW SULFUR	MMBTU
ECRGASM	CRUDE GAS FROM DESULFURIZER - MEDIUM SULFUR	MMBTU
ECTH2O	COOLING TOWER MAKEUP WATER	GAL
ECTNOBLD	NON-OILY COOLING TOWER BLOWDOWN	GAL
ECTTDS	TDS IN COOLING TOWER MAKEUP	LB X E-
EDMH2O	WATER TO DEMINERALIZER	GAL
EDMTDS	DISSOLVED SOLIDS TO DEMINERALIZER	LB X E-
EDMWCD	BOILER WATER-PROCESS WATER	GAL X 1
EEAH2O	WATER FOR DISCHARGE BACK TO RIVER	GAL
EEATDS	DISSOLVED SOLIDS DISCHARGE TOTAL	LB X E-
EELEC	ELECTRICITY REQUIREMENTS	KWH
EEVH2O	BRINE WATER TO EVAPORATOR	GAL
EEVTDS	SALTS TO EVAPORATOR	GAL
EGOSO2	SO2 FROM GAS/OIL PLANTS - OIL	LB X 1
EGTCTE	ELECTRICITY FROM GAS TURBINES IN CGCC	KWH
EGTCTS	SO2 FROM CGCC GAS TURBINES	LB X 1
EGTRES	RESTRICTION OF H2S TO CGCC GAS TURBINES	LB X 1
EHISCAV	HIGH-SULFUR COAL AVAILABILITY	MBTU
EHISOAV	HIGH-SULFUR OIL AVAILABILITY	MBTU
EHOTGAS	HOT GAS FROM CGCC GAS TURBINES	LB X E
EHTCONST	HEAT EFFLUENT CONSTRAINT	BTU X
ELCGCC	CGCC PLANT INVESTMENT LIMIT	CENTS
ELCONS	LIMIT ON INTERNAL PLANT CONSUMPTION OF WATER	GAL
ELEGO	CONSTRAINT ON OLD GAS PLANTS	KWH
ELIMESTN	LIMESTONE TO SCRUBBER	LB X 1
ELNATGAS	CONSTRAINT ON NATURAL GAS AVAILABILITY	MBTU
ELOCC	CONSTRAINT ON OLD COMBINED CYCLE PLANTS	KWH
ELOLDCL	CONSTRAINT ON OLD COAL PLANTS	KWH
ELOLDO	CONSTRAINT ON OLD OIL PLANTS	KWH
ELOSCAV	LOW SULFUR COAL AVAILABILITY	MBTU
ELOSOAV	LOW SULFUR OIL AVAILABILITY	MBTU
ELRIVERW	CONSTRAINT ON RIVER WATER USE	GAL
EMEDSCAV	MEDIUM SULFUR COAL AVAILABILITY	MBTU
ENATGAS	NATURAL GAS USE	MBTU
ENCPART	NEW COAL-PLANT PARTICULATES DISCHARGES	LB X
ENCSO2	NEW COAL-PLANT SULFUR DIOXIDE DISCHARGES	LB X

```
ENEWCON    TOTAL NEW CONSTRUCTION CONSTRAINT                     CENTS
ENH3PD     AMMONIA USE                                           LB
ENOPART    NEW OIL-PLANT PARTICULATES DISCHARGES                 LB X E-2
ENOSO2     NEW OIL-PLANT SO2 DISCHARGES                          LB X 10
EOCPART    OLD COAL-PLANT PARTICULATES DISCHARGES                LB X E-2
EOCSO2     SO2 FROM OIL COAL PLANT                               LB X 10
EOHFUEL    OIL OR GAS FOR HIGH-SULFUR GAS-OIL UNITS              MBTU
EOILBURN   OIL USE                                               MBTU
EOILPART   OLD OIL PLANT PARTICULATES DISCHARGES                 LB X E-2
EOLDCNT    LOWER LIMIT ON USE OF OLD POWER PLANTS                KWH
EOLDCNT2   UPPER LIMIT ON USE OF OLD POWER PLANTS                KWH
EOLFUEL    OIL OR GAS TO GAS/OIL UNITS                           MBTU
EOOSO2     SO2 FROM OLD OIL PLANT                                LB X 10
EPOGAS     FUEL TO GASIFIER                                      MBTU
ERIVERW    USE OF RIVER WATER                                    GAL X 100
ERNOX      NOX EFFLUENT                                          LB X E-2
ERPART     PARTICULATES FROM BURNING FUELS                       LB X E-2
ERSTAKH    STACK HEAT                                            MBTU
ESCRGUL    UPPER LIMIT ON WASTEWATER RECYCLE TO SCRUBBER         GAL
ESLDGWAT   DECANTED WATER FROM STACK GAS SCRUBBER SLUDGE         GAL X 100
SLUDGE     SLUDGE FROM SGS UNITS                                 LB
SOCLBD     SOLIDS IN CLARIFIER BLOWDOWN                          LB
SOLID      SOLIDS IN EFFLUENT                                    LB X E-1
SOLSAL     SOLIDIFIED SALT RESIDUE FROM EVAPORATOR               LB X E-2
SSLD       SUSPENDED SOLIDS FOR DISCHARGE                        LB
SULF       ELEMENTAL SULFUR                                      LB
SULPACD    SUPPLY OF H2SO4                                       LB
STEAM      MEDIUM PRESSURE STEAM                                 MBTU X 10
TOTHT      TOTAL HEAT TO COOLING                                 BTU X E4
TSOLID     ACCOUNTING ROW FOR SOLIDS                             LB
TXSO2      TAX ON SO2                                            LB X 10
ULBBLOR    UPPER LIMIT ON BOILER BLOWDOWN RECYCLE                GAL
ULPART     CONSTRAINT ON PARTICULATES DISCHARGE                  LB X E-2
ULSO2      CONSTRAINT ON SO2 DISCHARGE                           LB X 10
ULXCT      CONSTRAINT ON EXISTING COOLING TOWERS                 MBTU
ULXOT      CONSTRAINT ON EXISTING ONCE THROUGH COOLING           MBTU
XHCL       SUPPLY OF HCL                                         LB X E-2
WSTHT      HEAT TO COOLING                                       BTU X E4
WSTH84     WASTE HEAT AT 84 DEGREES                              MMBTU
WSTH90     WASTE HEAT AT 90 DEGREES                              MMBTU
WSTH137    WASTE HEAT AT 137 DEGREES                             MMBTU
ZCTPART    COOLING TOWER PARTICULATES                            LB X E-2
ZH2OOUT    WATER EFFLUENT                                        GAL X 100
ZOTH2O     ONCE THROUGH COOLING CONSTRAINT                       GAL
ZULDISL    CONSTRAINT ON DISCHARGE OF DISSOLVED SOLIDS           LB X 10
ZULSSLD    CONSTRAINT ON DISCHARGE OF SUSPENDED SOLIDS           LB
J          COST ROW                                              $
J2         NATURAL GAS COST CHANGE ROW                           $
JL         CHANGE IN PRICE OF LOW-SULFUR COAL                    $
JRD        DECREASE IN CAPITAL RATE OF RETURN                    $
JRU        INCREASE IN CAPITAL RATE OF RETURN                    $
JW         CHANGE IN WATER PRICE                                 $
```

Column Definitions

EAIRCOOL	AIR COOLING	MMBTU
EASHEV	ASH WATER TO EVAPORATOR	GAL
EASHEA	ASH WATER FOR DISCHARGE	FAL
EBBLOCT	BOILER BLOWDOWN TO COOLING TOWER	GAL
EBOBLOEV	BOILER BLOWDOWN TO EVAPORATOR	GAL
EBOBLOEA	BOILER BLOWDOWN FOR DISCHARGE	GAL
ECLAR	CLARIFIED WATER	MGAL
ECLNO	ONCE THROUGH COOLING	BTU X
ECOAL	COAL USE	MBTU
ECWH137	COOLING AT 137 DEGREES	MMBTU
ECWH84	COOLING AT 84 DEGREES	MMBTU
ECWH90	COOLING AT 90 DEGREES	MMBTU
EDESH	DESULFURIZATION OF CRUDE GAS FROM GASIFIER USING HIGH-SULFUR COAL	MMBTU
EDESIIH	DESULFURIZATION WITH TAIL GAS RECYCLE OF CRUDE GAS FROM GASIFIER USING HIGH-SULFUR COAL	MMBTU
EDESIIL	DESULFURIZATION WITH TAIL GAS RECYCLE OF CRUDE GAS FROM GASIFIER USING LOW-SULFUR COAL	MMBTU
EDESIIM	DESULFURIZATION WITH TAIL GAS RECYCLE OF CRUDE GAS FROM GASIFIER USING MEDIUM-SULFUR COAL	MMBTU
EDESL	DESULFURIZATION OF CRUDE GAS FROM GASIFIER USING LOW-SULFUR COAL	MMBTU
EDESM	DESULFURIZATION OF CRUDE GAS FROM GASIFIER USING MEDIUM-SULFUR COAL	MMBTU
EDMCLW	CLARIFIED WATER TO DEMINERALIZER	MGAL
EDMINH20	DEMINERALIZED WATER	MGAL
EDMNOBLD	DEMINERALIZER BLOWDOWN	GAL
EDMTDSCL	SOLIDS TO DEMINERALIZER USING HCL	.0095
EDMTDSSU	SOLIDS TO DEMINERALIZER USING SULFURIC ACID	LB X
EEACS	CLARIFIER SLUDGE FOR DISCHARGE TO RIVER	LB
EEANOBLD	BLOWOWN FOR DISHCARGE TO RIVER	GAL
EEH20	DISCHARGE OF WATER	GAL
EESOLID	DISCHARGE OF SOLIDS	LB
EESSLD	DISCHARGE OF SUSPENDED SOLIDS	LB
EETDS	DISCHARGE OF DISSOLVED SOLIDS	LB
EEVNOH20	EVAPORATOR WATER	GAL
EEVNOTDS	EVAPORATOR SALTS	LB X
EGASH	GASIFIER USING HIGH-SULFUR COAL	MMBTU
EGASL	GASIFIER USING LOW-SULFUR COAL	MMBTU
EGASM	GASSIFIER USING MEDIUM-SULFUR COAL	MMBTU
EGN	NEW ELECTRIC PLANT USING NATURAL GAS	KWH
EGO	OLD ELECTRIC PLANT USING NATURAL GAS	KWH
EGOSGS	STACK GAS SCRUBBING FOR OLD GAS-OR-OIL ELECTRIC PLANT	LB
EGOSO2T	UNTREATED SO2 EFFLUENT FROM OLD GAS-OR-OIL PLANT	LB X
EGTSO2CT	CGCC GAS TURBINE SO2 EFFLUENT	LB X
EGTELCT	CGCC GAS TURBINE ELECTRICAL OUTPUT	KWH
EGTELCT	CGCC GAS TURBINE BURNING CRUDE GAS FROM HIGH-SULFUR COAL	KWH
EGTWL	CGCC GAS TURBINE BURNING CRUDE GAS FROM LOW-SULFUR COAL	KWH
EGTWM	CGCC GAS TURBINE BURNING CRUDE GAS FROM MEDIUM-SULFUR COAL	KWH
EGTWOH	CGCC GAS TURBINE BURNING CLEANED GAS FROM HIGH-SULFUR COAL	KWH

WOL	CGCC GAS TURBINE BURNING CLEANED GAS FROM LOW-SULFUR COAL	KWH X 100
WOM	CGCC GAS TURBINE BURNING CLEANED GAS FROM MEDIUM-SULFUR COAL	KWH X 10
OH	OLD ELECTRIC PLANT BURNING GAS OR HIGH-SULFUR OIL	KWH
OL	OLD ELECTRIC PLANT BURNING GAS OR LOW-SULFUR OIL	KWH
SC	HIGH-SULFUR COAL SUPPLY	MBTU
SO	HIGH-SULFUR OIL SUPPLY	MBTU
CS	LANDFILL CALRIFIER SLUDGE	LB
SC	LOW-SULFUR COAL SUPPLY	MBTU
SO	LOW-SULFUR OIL SUPPLY	MBTU
DSC	MEDIUM-SULFUR COAL SUPPLY	MBTU
C	NEW COMBINED CYCLE POWER PLANT	KWH
PARS	UNTREATED PARTICULATES FROM NEW COAL-BURNING PLANTS	LB X 100
SGP	PARTICULATE PRECIPITATORS FOR NEW COAL-BURNING PLANTS	LB X 100
SGS	STACK GAS SCRUBBERS FOR NEW COAL-BURNING PLANTS	LB
SO2T	UNTREATED SO2 FROM NEW COAL BURNING PLANTS	LB X 10
	NEW COOLING TOWERS	MMBTU
CH	NEW POWER PLANTS USING HIGH-SULFUR COAL	KWH
CL	NEW POWER PLANTS USING LOW-SULFUR COAL	KWH
CM	NEW POWER PLANTS USING MEDIUM-SULFUR COAL	KWH
OH	NEW POWER PLANTS USING HIGH-SULFUR OIL	KWH
OL	NEW POWER PLANTS USING LOW-SULFUR OIL	KWH
ARS	UNTREATED PARTICULATES FROM NEW OIL-BURNING PLANTS	LB X 100
GP	PARTICULATE PRECIPITATORS FOR NEW OIL-BURNING PLANTS	LB X 100
GS	STACK GAS SCRUBBERS FOR NEW OIL-BURNING PLANTS	LB
O2T	UNTREATED SO2 FROM NEW OIL-BURNING PLANTS	LB X 10
SOH	NATURAL GAS FOR GAS-OR-HIGH-SULFUR OIL PLANTS	MBTU
SOL	NATURAL GAS FOR GAS-OR-LOW-SULFUR OIL PLANTS	MBTU
	OLD COMBINED CYCLE PLANTS	KWH
ARS	UNTREATED PARTICULATES FROM OLD COAL-BURNING PLANTS	LB X E-2
GP	PARTICULATE PRECIPITATORS FOR OLD COAL-BURNING PLANTS	LB X E-2
GS	STACK GAS SCRUBBERS FOR OLD COAL-BURNING PLANTS	LB X 10
O2T	UNTREATED SO2 FROM OLD COAL-BURNING PLANTS	LB X 10
SO	HIGH-SULFUR OIL FOR GAS-OR-HIGH-SULFUR OIL PLANTS	MBTU
PARS	UNTREATED PARTICULATES FROM OIL-BURNING PLANTS	LB X E-2
SGP	PARTICULATE PRECIPITATORS FOR OLD OIL-BURNING PLANTS	LB X E-2
CH	OLD COAL-BURNING PLANT USING HIGH-SULFUR COAL	KWH
CL	OLD COAL-BURNING PLANT USING LOW-SULFUR COAL	KWH
CM	OLD COAL-BURNING PLANT USING MEDIUM-SULFUR COAL	KWH
CH	OLD COAL-BURNING PLANT USING HIGH-SULFUR COAL	KWH
OH	OLD OIL-BURNING PLANT USING HIGH-SULFUR OIL	KWH
OL	OLD OIL-BURNING PLANT USING LOW-SULFUR OIL	KWH
SO	LOW-SULFUR OIL FOR PLANTS BURNING GAS OR LOW-SULFUR OIL	MBTU
	STACK GAS SCRUBBERS FOR OLD OIL-BURNING PLANTS	LB
2T	UNTREATED SO2 FROM OLD OIL-BURNING PLANTS	LB X 10
TAX	PARTICULATES EFFLUENT TAX	LB X E-2
	EXISTING COOLING TOWERS	MMBTU
W	COOLING TOWER WITH CLARIFIED WATER MAKEUP	GAL
M	COOLING TOWER WITH DEMINERALIZED WATER MAKEUP	GAL
DS	COOLING TOWER TOTAL DISSOLVED SOLIDS	LB X E-2
W	RIVER WATER	GAL
AX	SO2 EFFLUENT TAX	LB X 10
	STEAM ELECTRIC PART OF CGCC	KWH X 100
H	STACK HEAT	MMBTU
TX	HEAT-TO-WATER EFFLUENT TAX	BTU X E4
S	WATER CONSUMPTION	GAL

EXCLGASH	TRANSFER OF ¤ECLGASH¤ TO ¤EPOGAS¤	MMBTU
EXCLGASL	TRANSFER OF ¤ECLGASL¤ TO ¤EPOGAS¤	MMBTU
EXCLGASM	TRANSFER OF ¤ECLGASM¤ TO ¤EPOGAS¤	MMBTU
EXCLAR	DECANTED WATER FOR PRECIPITATORS	GAL X
EXLOSO	TRANSFER OF LOW-SULFUR OIL TO ¤EPOGAS¤	MBTU
EXNATGS	TRANSFER OF NATURAL GAS TO ¤EXPOGAS¤	MBTU
EXRIV	RIVER WATER TO PRECIPITATORS	GAL X
EZALUM	ALUMINUM SULFATE HYDRATE SUPPLY	LB
EZLIMEST	LIMESTONE SUPPLY	LB X 1
EZNATGAS	NATURAL GAS SUPPLY	MBTU
EZNH3PD	AMMONIA SUPPLY	LB
EZNOX	NOX EFFLUENT	LB X E
EZOIL	OIL USE	MBTU
EZSLUDGE	SLUDGE DISPOSAL	LB
EZSOCBLD	CLARIFIER BLOWDOWN SOLIDS	LB
EZSULF	SULFUR	LB
EZSTEAM	STEAM PRODUCTION	MMBTU
EZSUACD	SULFURIC ACID SUPPLY	LB
EZDNAOH	CAUSTIC SUPPLY	LB
EZZHCL	HYDROCHLORIC ACID SUPPLY	LB X

Data Matrix

```
ROWS
 L  EAIRHT
 E  EADNAOH
 E  EALUM
 E  EASHWEV
 E  EBHISC
 E  EBHISO
 E  EBLOSC
 E  EBLOSO
 E  EBOILBLO
 E  EBMEDSC
 E  ECLARW
 G  ECLGASH
 G  ECLGASL
 G  ECLGASM
 L  ECLPART
 L  ECLSO2
 E  ECLWATER
 E  ECOALBRN
 E  ECONS
 G  ECRGASH
 G  ECRGASL
 G  ECRGASM
 E  ECTH2O
 E  ECTNOBLD
 E  ECTTDS
 E  EDMH2O
 E  EDMTDS
 E  EDMWCD
 E  EEAH2O
 E  EEATDS
 G  EELEC
 E  EEVH2O
 E  EEVTDS
 E  EGOSO2
 E  EGTCTE
 E  EGTCTS
 G  EGTRES
 L  EHISCAV
 L  EHISOAV
 G  EHOTGAS
 L  EHTCONST
 L  ELCGCC
 L  ELCONS
 L  ELEGO
 E  ELIMESTN
 L  ELNATGAS
 L  ELOCC
 L  ELOLDCL
 L  ELOLDC
 L  ELOSCAV
 L  ELOSOAV
 L  ELRIVERW
 L  EMEDSCAV
 E  ENATGAS
 E  ENCPART
 E  ENCSO2
 L  ENEWCON
 E  ENH3PD
 E  ENOPART
 E  ENOSO2
 E  EOCPART
 E  EOCSO2
 E  EOHFUEL
 E  EOILBURN
 E  EOILPART
 G  EOLDCNT
 L  EOLDCNT2
 E  EOLFUEL
 E  EOOSO2
 E  EPOGAS
 E  ERIVERW
 E  ERNOX
 E  ERPART
 E  ERSTAKH
 L  ESCRGUL
 E  ESLDGWAT
 E  ESLUDGE
 E  ESOCLBD
 E  ESOLID
 N  ESOLSAL
 E  ESSLD
 E  ESULF
 E  ESULPACD
 E  ESTEAM
 E  ETOTHT
 N  ETSOLID
 E  ETXSO2
 L  EULBBLOR
 L  EULPART
 L  EULSO2
 E  EULXCT
 L  EULXOT
 E  EXHCL
 E  EWSTHT
 E  EWSTH84
 E  EWSTH90
 E  EWSTH137
 N  EZCTPART
 L  EZH2OOUT
 L  EZOTH2O
 L  EZULDISL
 L  EZULSSLD
 N  OBJ
 N  OBJL
 N  OBJRD
 N  OBJRU
 N  OBJW
 N  OBJ2

COLUMNS
    EAIRCOOL   EELEC        -7.79     ENEWCON      31.5
    EAIRCOOL   EWSTH137      1.0      OBJ           0.0591
    EAIRCOOL   OBJRD         -.033    OBJRU         0.033
    EASHEA     EASHWEV      -1.0      EEAH2O        1.0
    EASHEA     EEATDS        4.2
    EASHEV     EASHWEV      -1.0      EEVH2O        1.0
    EASHEV     EEVTDS        4.2
    EBBLOCT    EBOILBLO     -1.0      ECTH2O        1.0
    EBBLOCT    ECTTDS        0.042    EULBBLOR      1.0
    EBOBLOEA   EBOILBLO     -1.0      EEAH2O        1.0
    EBOBLOEA   EEATDS        0.042
    EBOBLOEV   EBOILBLO     -1.0      EEVH2O        1.0
    EBOBLOEV   EEVTDS        0.042
    ECLAR      EALUM         -.47     ECLARW      100.0
```

ECLAR	EELEC	-.145	ERIVERW	-10.5
ECLAR	ESOCLBD	1.0	OBJ	0.32
ECLAR	OBJRD	-.18	OBJRU	0.18
ECLNO	EEAH2O	78.2	EEATDS	7.8
ECLNO	EELEC	-0.0102	ERIVERW	-0.782
ECLNO	EULXOT	1.0	EWSTH84	0.01
ECLNO	EZOTH2O	78.2		
ECOAL	ECOALBRN	-1.0		
ECWH137	EELEC	-24.05	EWSTHT	-100.0
ECWH137	EWSTH137	-1.0		
ECWH84	EELEC	3.5	EWSTHT	-100.0
ECWH84	EWSTH84	-1.0		
ECWH90	EELEC	-0.5	EWSTHT	-100.0
ECWH90	EWSTH90	-1.0		
EDESH	ECLGASH	1.0	ECRGASH	-1.006
EDESH	EDMWCD	-.01	EELEC	-1.234
EDESH	ELCGCC	52.80	ENEWCON	52.80
EDESH	ESTEAM	-1.068	ESULF	6.47
EDESH	ETXSO2	0.068	EULSO2	0.068
EDESH	EWSTHT	0.576	OBJ	0.111
EDESH	OBJRD	-.074	OBJRU	0.074
EDESIIH	ECLGASH	1.00	ECRGASH	-1.006
EDESIIH	EDMWCD	-.01	EELEC	-1.234
EDESIIH	ELCGCC	55.60	ENEWCON	55.60
EDESIIH	ESTEAM	-1.068	ESULF	6.80
EDESIIH	EWSTHT	0.576	OBJ	0.116
EDESIIH	OBJRD	-.078	OBJRU	0.078
EDESIIL	ECLGASL	1.00	ECRGASL	-1.005
EDESIIL	EDMWCD	-.01	EELEC	-1.234
EDESIIL	ELCGCC	55.60	ENEWCON	55.60
EDESIIL	ESTEAM	-1.068	ESULF	0.847
EDESIIL	EWSTHT	0.576	OBJ	0.116
EDESIIL	OBJRD	-.078	OBJRU	0.078
EDESIIM	ECLGASM	1.00	ECRGASM	-1.0054
EDESIIM	EDMWCD	-.01	EELEC	-1.234
EDESIIM	ELCGCC	55.60	ENEWCON	55.60
EDESIIM	ESTEAM	-1.068	ESULF	2.72
EDESIIM	EWSTHT	0.576	OBJ	0.116
EDESIIM	OBJRD	-.078	OBJRU	0.078
EDESL	ECLGASL	1.0	ECRGASL	-1.005
EDESL	EDMWCD	-.01	EELEC	-.034
EDESL	ELCGCC	52.80	ENEWCON	52.80
EDESL	ESTEAM	-1.068	ESULF	0.81
EDESL	ETXSO2	0.0085	EULSO2	0.0085
EDESL	EWSTHT	0.576	OBJ	0.111
EDESL	OBJRD	-.074	OBJRU	0.074
EDESM	ECLGASM	1.0	ECRGASM	-1.0054
EDESM	EDMWCD	-.01	EELEC	-.034
EDESM	ELCGCC	52.80	ENEWCON	52.80
EDESM	ESTEAM	-1.1068	ESULF	2.59
EDESM	ETXSO2	0.027	EULSO2	0.027
EDESM	EWSTHT	0.576	OBJ	0.111
EDESM	OBJRD	-.074	OBJRU	0.074
EDMCLW	ECLARW	-100.0	EDMH2O	1000.0
EDMCLW	EDMTDS	100.0		
EDMINH2O	EDMH2O	-1000.0	EDMWCD	100.0
EDMINH2O	OBJ	0.3	OBJRD	-.168
EDMINH2O	OBJRU	0.168		
EDMNOBLD	ECTNOBLD	-1.0	EDMH2O	1.0

EDMNOBLD	EDMTDS	1.46		
EDMTDSCL	EADNAOH	-0.0062	ECLARW	-0.027
EDMTDSCL	EDMH2O	0.19	EDMTDS	-0.956
EDMTDSCL	EEVH2O	0.084	EEVTDS	1.78
EDMTDSCL	EXHCL	-0.57		
EDMTDSSU	EADNAOH	-0.0062	ECLARW	-0.03
EDMTDSSU	EDMTDS	-1.0	EEAH2O	0.3
EDMTDSSU	EEATDS	2.0	ESULPACD	-0.0076
EEACS	EEAH2O	20.0	EEATDS	5.83
EEACS	ESOCLBD	-1.0	ESSLD	1.0
EEANOBLD	ECTNOBLD	-1.0	EEAH2O	1.0
EEANOBLD	EEATDS	1.466		
EEH2O	ECONS	-1.0	EEAH2O	-1.0
EEH2O	EZH2OOUT	0.1		
EESOLID	ESOLID	-10.0	ETSOLID	1.0
EESOLID	OBJ	0.1	OBJRD	-.056
EESOLID	OBJRU	0.056		
EESSLD	ESSLD	-1.0	EZULSSLD	1.0
EETDS	EEATDS	-1.0	EZULDISL	0.1
EEVNOH2O	EDMH2O	1.0	EDMTDS	0.0833
EEVNOH2O	EDMWCD	.022	EELEC	-0.03
EEVNOH2O	EEVH2O	-1.0	OBJ	0.004
EEVNOH2O	OBJRD	-.0022	OBJRU	0.0022
EEVNOTDS	EEVTDS	-1.0	ESOLSAL	1.0
EGASH	EBHISC	-1166.1	ECOALBRN	1166.1
EGASH	ECRGASH	1.0	EDMWCD	-0.709
EGASH	EELEC	-10.01	ELCGCC	144.60
EGASH	ENEWCON	144.60	EPOGAS	-34.0
EGASH	ESOLID	115.517	OBJ	0.362
EGASH	OBJRD	-.203	OBJRU	0.203
EGASL	EBLOSC	-1166.1	ECOALBRN	1166.1
EGASL	ECRGASL	1.0	EDMWCD	-0.709
EGASL	EELEC	-10.01	ELCGCC	144.60
EGASL	ENEWCON	144.60	EPOGAS	-34.0
EGASL	ESOLID	115.517	OBJ	0.362
EGASL	OBJRD	-.203	OBJRU	0.203
EGASM	EBMEDSC	-1166.1	ECOALBRN	1166.1
EGASM	ECRGASM	1.0	EDMWCD	-0.709
EGASM	EELEC	-10.01	ELCGCC	144.60
EGASM	ENEWCON	144.60	EPOGAS	-34.0
EGASM	ESOLID	115.517	OBJ	0.362
EGASM	OBJRD	-.203	OBJRU	0.203
EGN	EBOILBLO	0.007	EDMWCD	-.0007
EGN	EELEC	1.0	ENATGAS	-8.5
EGN	ENEWCON	3.3	ERSTAKH	1.0
EGN	ETOTHT	0.40	EWSTHT	0.40
EGN	OBJ	0.0060	OBJRD	-.0034
EGN	OBJRU	0.0034		
EGO	EBOILBLO	0.01	EDMWCD	-.001
EGO	EELEC	1.0	ELEGO	1.0
EGO	ENATGAS	-10.84	ENEWCON	0.0
EGO	EOLDCNT	1.0	EOLDCNT2	1.0
EGO	ERSTAKH	1.44	ETOTHT	0.56
EGO	EWSTHT	0.56	OBJ	0.0053
EGOSGS	EELEC	-7.4	EGOSO2	-1.0
EGOSGS	ELIMESTN	-2.46	ENATGAS	-53.57
EGOSGS	ENEWCON	1078.6	ENH3PD	-0.012
EGOSGS	ERIVERW	-.270	ESLUDGE	58.67
EGOSGS	ETXSO2	0.1	EULSO2	0.1

EGOSGS	OBJ	1.95	OBJRD	-1.09
EGOSGS	OBJRU	1.09		
EGOSO2T	ECLSO2	1.0	EGOSO2	-1.0
EGOSO2T	ETXSO2	1.0	EULSO2	1.0
EGTELCT	EGTCTE	-1.0	EGTRES	1.0
EGTSO2CT	EGTCTS	-1.0	EGTRES	-100.0
EGTWH	ECLSO2	1.542	ECRGASH	-1.228684
EGTWH	EELEC	100.0	EGTCTE	100.0
EGTWH	EGTCTS	1.8735	EHOTGAS	2.8920
EGTWH	ELCGCC	246.7	ENEWCON	246.7
EGTWH	ERNOX	0.01	ETXSO2	1.8735
EGTWH	EULSO2	1.8735	OBJ	0.520
EGTWH	OBJRD	-.29	OBJRU	0.29
EGTWL	ECLSO2	0.1928	ECRGASL	-1.228684
EGTWL	EELEC	100.0	EGTCTE	100.0
EGTWL	EGTCTS	0.2342	EHOTGAS	2.8920
EGTWL	ELCGCC	246.7	ENEWCON	246.7
EGTWL	ERNOX	0.01	ETXSO2	0.2342
EGTWL	EULSO2	0.2342	OBJ	0.520
EGTWL	OBJRD	-.29	OBJRU	0.29
EGTWM	ECLSO2	0.6168	ECRGASM	-1.228684
EGTWM	EELEC	100.0	EGTCTE	100.0
EGTWM	EGTCTS	0.7494	EHOTGAS	2.8920
EGTWM	ELCGCC	246.7	ENEWCON	246.7
EGTWM	ERNOX	0.01	ETXSO2	0.7494
EGTWM	EULSO2	0.7494	OBJ	0.520
EGTWM	OBJRD	-.29	OBJRU	0.29
EGTWOH	ECLGASH	-1.228684	EELEC	100.0
EGTWOH	EGTCTE	100.0	EGTCTS	0.3315
EGTWOH	EHOTGAS	2.8490	ELCGCC	246.7
EGTWOH	ENEWCON	246.7	ERNOX	0.01
EGTWOH	ETXSO2	0.3315	EULSO2	0.3315
EGTWOH	OBJ	0.520	OBJRD	-.29
EGTWOH	OBJRU	0.29		
EGTWOL	ECLGASL	-1.228684	EELEC	100.0
EGTWOL	EGTCTE	100.0	EGTCTS	0.04144
EGTWOL	EHOTGAS	2.8490	ELCGCC	246.7
EGTWOL	ENEWCON	246.7	ERNOX	0.01
EGTWOL	ETXSO2	0.04144	EULSO2	0.04144
EGTWOL	OBJ	0.520	OBJRD	-.29
EGTWOL	OBJRU	0.29		
EGTWOM	ECLGASM	-1.228684	EELEC	100.0
EGTWOM	EGTCTE	100.0	EGTCTS	0.1326
EGTWOM	EHOTGAS	2.8490	ELCGCC	246.7
EGTWOM	ENEWCON	246.7	ERNOX	0.01
EGTWOM	ETXSO2	0.1326	EULSO2	0.1326
EGTWOM	OBJ	0.520	OBJRD	-.29
EGTWOM	OBJRU	0.29		
EGWOH	EBOILBLO	0.01	EDMWCD	-.001
EGWOH	EELEC	1.0	ELEGO	1.0
EGWOH	ENEWCON	0.017	EOHFUEL	-10.86
EGWOH	EOLDCNT	1.0	EOLDCNT2	1.0
EGWOH	ERSTAKH	1.48	ETOTHT	0.57
EGWOH	EWSTHT	0.57	OBJ	0.0055
EGWOL	EBOILBLO	0.01	EDMWCD	-.001
EGWOL	EELEC	1.0	ELEGO	1.0
EGWOL	ENEWCON	0.017	EOLDCNT	1.0
EGWOL	EOLDCNT2	1.0	EOLFUEL	-10.86

EGWOL	ERSTAKH	1.29	ETOTHT	0.57
EGWOL	EWSTHT	0.57	OBJ	0.0055
EHISC	EBHISC	1.0	EHISCAV	1.0
EHISC	OBJ	.44E-3		
EHISO	EBHISO	1.0	EHISOAV	1.0
EHISO	EOILBURN	1.0	OBJ	1.19E-3
ELFCS	ECLARW	1.96	EELEC	-.0255
ELFCS	ESOCLBD	-1.0	ETSOLID	1.0
ELFCS	OBJ	0.033	OBJRD	-.018
ELFCS	OBJRU	0.018		
ELOSC	EBLOSC	1.0	ELOSCAV	1.0
ELOSC	OBJ	0.61E-3	OBJL	25.00E-3
ELOSO	EBLOSO	1.0	ELOSOAV	1.0
ELOSO	EOILBURN	1.0	OBJ	1.39E-3
EMEDSC	EBMEDSC	1.0	EMEDSCAV	1.0
EMEDSC	OBJ	0.50E-3		
ENCC	EBOILBLO	0.004	EDMWCD	-.0004
ENCC	EELEC	1.0	ENATGAS	-7.0
ENCC	ENEWCON	3.0	ERSTAKH	0.6
ENCC	ETOTHT	0.29	EWSTHT	0.29
ENCC	OBJ	0.0057	OBJRD	-.003
ENCC	OBJRU	0.003		
ENCPARS	ECLPART	1.0	ENCPART	-1.0
ENCPARS	ERPART	1.0	EULPART	1.0
ENCSGP	EASHWEV	0.2	ECLWATER	-0.08
ENCSGP	EELEC	-1.30E-6	ENCPART	-30.0
ENCSGP	ENEWCON	0.247	ERPART	1.0
ENCSGP	ESLDGWAT	0.078	ESLUDGE	0.4
ENCSGP	EULPART	1.0	OBJ	0.0052
ENCSGP	OBJRD	-.003	OBJRU	0.003
ENCSGS	EELEC	-7.4	ELIMESTN	-2.46
ENCSGS	ENATGAS	-53.57	ENCSO2	-1.0
ENCSGS	ENEWCON	108.4	ENH3PD	-0.012
ENCSGS	ERIVERW	-.270	ESLUDGE	58.67
ENCSGS	ETXSO2	0.1	EULSO2	0.1
ENCSGS	OBJ	0.45	OBJRD	-.25
ENCSGS	OBJRU	0.25		
ENCSO2T	ECLSO2	1.0	ENCSO2	-1.0
ENCSO2T	ETXSO2	1.0	EULSO2	1.0
ENCT	ECTH2O	-120.0	EELEC	-1.875
ENCT	ENEWCON	24.2	EWSTH90	1.0
ENCT	EZCTPART	2.5	OBJ	.033
ENCT	OBJRD	-.018	OBJRU	0.018
ENEWCH	EBHISC	-8.6	EBOILBLO	0.007
ENEWCH	ECOALBRN	8.6	EDMWCD	-.0007
ENEWCH	EELEC	1.0	ENCPART	11.55
ENEWCH	ENCSO2	0.01032	ENEWCON	3.83
ENEWCH	ERNOX	2.46	ERSTAKH	1.0
ENEWCH	ETOTHT	0.41	EWSTHT	0.41
ENEWCH	OBJ	0.0063	OBJRD	-.0035
ENEWCH	OBJRU	0.0035		
ENEWCL	EBLOSC	-8.6	EBOILBLO	0.007
ENEWCL	ECOALBRN	8.6	EDMWCD	-.0007
ENEWCL	EELEC	1.0	ENCPART	11.55
ENEWCL	ENCSO2	0.00129	ENEWCON	3.83
ENEWCL	ERNOX	2.46	ERSTAKH	1.0
ENEWCL	ETOTHT	0.41	EWSTHT	0.41
ENEWCL	OBJ	0.0063	OBJRD	-.0035

ENEWCL	OBJRU	0.0035		
ENEWCM	EBMEDSC	-8.6	EBOILBLO	0.007
ENEWCM	ECOALBRN	8.6	EDMWCD	-.0007
ENEWCM	EELEC	1.0	ENCPART	11.55
ENEWCM	ENCSO2	0.00413	ENEWCON	3.83
ENEWCM	ERNOX	2.46	ERSTAKH	1.0
ENEWCM	ETOTHT	0.41	EWSTHT	0.41
ENEWCM	OBJ	0.0063	OBJRD	-.0035
ENEWCM	OBJRU	0.0035		
ENEWOH	EBHISO	-8.5	EBOILBLO	0.01
ENEWOH	EDMWCD	-.001	EELEC	1.0
ENEWOH	ENEWCON	3.8	ENOPART	0.11
ENEWOH	ENOSO2	0.0019	ERNOX	1.61
ENEWOH	ERSTAKH	1.0	ETOTHT	0.4
ENEWOH	EWSTHT	0.4	OBJ	0.0061
ENEWOH	OBJRD	-.0034	OBJRU	0.0034
ENEWOL	EBLOSO	-8.5	EBOILBLO	0.01
ENEWOL	EDMWCD	-.001	EELEC	1.0
ENEWOL	ENEWCON	3.8	ENOPART	0.11
ENEWOL	ENOSO2	0.0005	ERNOX	1.61
ENEWOL	ERSTAKH	1.0	ETOTHT	0.4
ENEWOL	EWSTHT	0.4	OBJ	0.0061
ENEWOL	OBJRD	-.0034	OBJRU	0.0034
ENOPARS	ECLPART	1.0	ENOPART	-1.0
ENOPARS	ERPART	1.0	EULPART	1.0
ENOSGP	EASHWEV	0.2	ECLWATER	-.08
ENOSGP	EELEC	-.001153	ENEWCON	26.1
ENOSGP	ENOPART	-30.0	ERPART	1.0
ENOSGP	ESLDGWAT	0.078	ESLUDGE	0.4
ENOSGP	EULPART	1.0	OBJ	0.54
ENOSGP	OBJRD	-.3	OBJRU	0.3
ENOSGS	EELEC	-7.4	ELIMESTN	-2.46
ENOSGS	ENATGAS	-53.57	ENEWCON	1527.0
ENOSGS	ENH3PD	-0.012	ENOSO2	-1.0
ENOSGS	ERIVERW	-.270	ESLUDGE	58.67
ENOSGS	ETXSO2	0.1	EULSO2	0.1
ENOSGS	OBJ	6.6	OBJRD	-3.7
ENOSGS	OBJRU	3.7		
ENOSO2T	ECLSO2	1.0	ENOSO2	-1.0
ENOSO2T	ETXSO2	1.0	EULSO2	1.0
ENTGSOH	ENATGAS	-1.0	EOHFUEL	1.0
ENTGSOL	ENATGAS	-1.0	EOLFUEL	1.0
EOCC	EBOILBLO	0.005	EDMWCD	-.0005
EOCC	EELEC	1.0	ELOCC	1.0
EOCC	ENATGAS	-8.0	EOLDCNT	1.0
EOCC	EOLDCNT2	1.0	ERSTAKH	0.7
EOCC	ETOTHT	0.35	EWSTHT	0.35
EOCC	OBJ	0.0051		
EOCPARS	ECLPART	1.0	EOCPART	-1.0
EOCPARS	ERPART	1.0	EULPART	1.0
EOCSGP	EASHWEV	0.2	ECLWATER	-0.08
EOCSGP	EELEC	-0.98E-6	EOCPART	-30.0
EOCSGP	ERPART	1.0	ESLDGWAT	0.078
EOCSGP	ESLUDGE	0.4	EULPART	1.0
EOCSGP	OBJ	0.004		
EOCSGS	EELEC	-7.4	ELIMESTN	-2.46
EOCSGS	ENATGAS	-53.57	ENEWCON	370.0
EOCSGS	ENH3PD	-0.012	EOCSO2	-1.0

EOCSGS	ERIVERW	-.270	ESLUDGE	58.67
EOCSGS	ETXSO2	0.1	EULSO2	0.1
EOCSGS	OBJ	0.66	OBJRD	-.37
EOCSGS	OBJRU	0.37		
EOCSO2T	ECLSO2	1.0	EOCSO2	-1.0
EOCSO2T	ETXSO2	1.0	EULSO2	1.0
EOHISO	EBHISO	-1.0	EGOSO2	0.00022
EOHISO	EOHFUEL	1.0	EOILPART	0.013
EOHISO	ERNOX	0.189		
EOILPARS	ECLPART	1.0	EOILPART	-1.0
EOILPARS	ERPART	1.0	EULPART	1.0
EOILSGP	EASHWEV	0.2	ECLWATER	-.08
EOILSGP	EELEC	-.001153	EOILPART	-30.0
EOILSGP	ERPART	1.0	ESLDGWAT	0.078
EOILSGP	ESLUDGE	0.4	EULPART	1.0
EOILSGP	OBJ	0.43		
EOLDCH	EBHISC	-10.25	EBOILBLO	0.01
EOLDCH	ECOALBRN	10.25	EDMWCD	-.001
EOLDCH	EELEC	1.0	ELOLDCL	1.0
EOLDCH	EOCPART	15.32	EOCSO2	.01368
EOLDCH	EOLDCNT	1.0	EOLDCNT2	1.0
EOLDCH	ERNOX	2.93	ERSTAKH	1.2
EOLDCH	ETOTHT	0.54	EWSTHT	.54
EOLDCH	OBJ	0.0059		
EOLDCL	EBLOSC	-10.25	EBOILBLO	0.01
EOLDCL	ECOALBRN	10.25	EDMWCD	-.001
EOLDCL	EELEC	1.0	ELOLDCL	1.0
EOLDCL	EOCPART	15.32	EOCSO2	0.00171
EOLDCL	EOLDCNT	1.0	EOLDCNT2	1.0
EOLDCL	ERNOX	2.93	ERSTAKH	1.2
EOLDCL	ETOTHT	0.54	EWSTHT	0.54
EOLDCL	OBJ	.0059		
EOLDCM	EBMEDSC	-10.25	EBOILBLO	0.01
EOLDCM	ECOALBRN	10.25	EDMWCD	-.001
EOLDCM	EELEC	1.0	ELOLDCL	1.0
EOLDCM	EOCPART	15.32	EOCSO2	0.00547
EOLDCM	EOLDCNT	1.0	EOLDCNT2	1.0
EOLDCM	ERNOX	2.93	ERSTAKH	1.2
EOLDCM	ETOTHT	.54	EWSTHT	0.54
EOLDCM	OBJ	.0059		
EOLDOH	EBHISO	-10.88	EBOILBLO	0.01
EOLDOH	EDMWCD	-.001	EELEC	1.0
EOLDOH	ELOLDO	1.0	EOILPART	0.14
EOLDOH	EOLDCNT	1.0	EOLDCNT2	1.0
EOLDOH	EOOSO2	0.0024	ERNOX	2.06
EOLDOH	ERSTAKH	1.48	ETOTHT	0.57
EOLDOH	EWSTHT	0.57	OBJ	0.0056
EOLDOL	EBLOSO	-10.88	EBOILBLO	0.01
EOLDOL	EDMWCD	-.001	EELEC	1.0
EOLDOL	ELOLDO	1.0	EOILPART	0.14
EOLDOL	EOLDCNT	1.0	EOLDCNT2	1.0
EOLDOL	EOOSO2	0.0006	ERNOX	2.06
EOLDOL	ERSTAKH	1.48	ETOTHT	0.57
EOLDOL	EWSTHT	0.57	OBJ	0.0056
EOLOSO	EBLOSO	-1.0	EGOSO2	0.00006
EOLOSO	EOILPART	0.013	EOLFUEL	1.0
EOLOSO	ERNOX	0.189		
EOOGS	EELEC	-7.4	ELIMESTN	-2.46

EOOGS	ENATGAS	-53.57	ENEWCON	2672.0
EOOGS	ENH3PD	-0.012	EOOSO2	-1.0
EOOGS	ERIVERW	-.270	ESLUDGE	58.67
EOOGS	ETXSO2	0.1	EULSO2	0.1
EOOGS	OBJ	4.95	OBJRD	-2.77
EOOGS	OBJRU	2.77		
EOOSO2T	ECLSO2	1.0	EOOSO2	-1.0
EOOSO2T	ETXSO2	1.0	EULSO2	1.0
EPARTAX	ERPART	-1.0	OBJ	0.00
ERCT	ECTH2O	-120.0	EELEC	-1.875
ERCT	EULXCT	1.0	EWSTH90	1.0
ERCT	EZCTPART	2.5	OBJ	.033
ERCTCW	ECLARW	-0.1	ECTH2O	1.0
ERCTCW	ECTTDS	0.10	EULBBLOR	-0.017
ERCTDM	ECTH2O	1.0	EDMWCD	-0.1
ERCTTDS	ECTH2O	-0.686	ECTNOBLD	0.686
ERCTTDS	ECTTDS	-1.0		
ERIVWW	ECONS	1.0	ELRIVERW	1.0
ERIVWW	ERIVERW	0.01	OBJ	0.0E-5
ERIVWW	OBJW	1.00		
ESO2TAX	ETXSO2	-1.0	OBJ	0.00
EST	EBOILBLO	1.32	EDMWCD	-.132
EST	EELEC	100.0	EHOTGAS	-3.827
EST	ELCGCC	394.6	ENEWCON	394.6
EST	ERSTAKH	00.222	ETOTHT	55.0
EST	EWSTHT	55.0	OBJ	0.990
EST	OBJRD	-.55	OBJRU	0.55
ESTAKH	EAIRHT	1.0	ERSTAKH	-1.0
ETOTHTX	EHTCONST	1.0	ETOTHT	-1.0
ETOTHTX	OBJ	0.000		
EWCONS	ECONS	-1.0	ELCONS	1.0
EXCLAR	ECLWATER	1.0	ESLDGWAT	-1.0
EXCLGASH	ECLGASH	-1.0	EPOGAS	1 000.0
EXCLGASH	ETXSO2	0.270	EULSO2	0.270
EXCLGASL	ECLGASL	-1.0	EPOGAS	1 000.0
EXCLGASL	ETXSO2	0.034	EULSO2	0.034
EXCLGASM	ECLGASM	-1.0	EPOGAS	1 000.0
EXCLGASM	ETXSO2	0.108	EULSO2	0.108
EXLOSO	EBLOSO	-1.0	EPOGAS	1.0
EXLOSO	ERNOX	0.189	ERPART	0.013
EXLOSO	EULPART	0.013	EULSO2	0.00006
EXNATGS	ENATGAS	-1.0	EPOGAS	1.0
EXRIV	ECLWATER	1.0	ERIVERW	-1.0
EZALUM	EALUM	1.0		
EZDNAOH	EADNAOH	1.0	OBJ	0.045
EZLIMEST	ELIMESTN	1.0	OBJ	0.025
EZNATGAS	ELNATGAS	1000.0	ENATGAS	1000.0
EZNATGAS	ERNOX	149.6	ERPART	4.0
EZNATGAS	EULPART	4.0	OBJ	0.64
EZNATGAS	OBJ2	1.40		
EZNH3PD	ENH3PD	1.0	OBJ	0.025
EZNOX	ERNOX	-1.0		
EZOIL	EOILBURN	-1.0		
EZSLUDGE	ESLUDGE	-1.0	ETSOLID	0.4
EZSLUDGE	OBJ	0.005	OBJRD	-.003
EZSLUDGE	OBJRU	0.003		
EZSOCBLD	ESOCLBD	-1.0		
EZSTEAM	EDMWCD	-8.8	EELEC	-1.98

```
    EZSTEAM    ELCGCC        3.6          ENEWCON       3.6
    EZSTEAM    EPOGAS     -960.0          ESTEAM      100.0
    EZSTEAM    OBJ           0.133        OBJRD         -.074
    EZSTEAM    OBJRU         0.074
    EZSUACD    ESULPACD      1.0          OBJ           0.0039
    EZSULF     ESULF        -1.0          OBJ           0.0
    EZZHCL     EXHCL      1000.0          OBJ           0.3
    RHSEL      EELEC      1.80E12         EULXOT      7.65E11
    RHSELC     EULSO2     2.16E9
    RHSF       EAIRHT     4.0E14          ECLPART     0.0E16
    RHSF       ECLSO2     1.0E15          EELEC       1.50E12
    RHSF       EHISCAV    6.0E15          EHISOAV     1.5E12
    RHSF       EHTCONST   1.20E15         ELCGCC      2.0E13
    RHSF       ELCONS     4.1E13          ELEGO       3.204E11
    RHSF       ELNATGAS   5.7E12          ELOCC       1.0E11
    RHSF       ELOLDCL    13.75E11        ELOLDO      2.923E11
    RHSF       ELOSCAV    6.16E12         ELOSOAV     0.2E12
    RHSF       ELRIVERW   3.12E14         EMEDSCAV    6.0E12
    RHSF       ENEWCON    2.0E13          EOLDCNT     1.375E12
    RHSF       EOLDCNT2   1.444E12        EULPART     1.75E16
    RHSF       EULSO2     5.2E15          EULXCT      1.19E9
    RHSF       EULXOT     6.373E11        EZH2OOUT       1.0
    RHSF       EZOTH2O    0.0E15          EZULDISL    0.0E15
    RHSF       EZULSSLD      1.0
    RHSNEW     ENEWCON    -1.0E13
    RHSPART    EULPART    -5.25E11
    RHSRIV     ELRIVERW   -3.12E14
    RHSSO2     EULSO2     -8.5E9
    RHSSULF    ECLSO2     -1.66E10
    RHSWAT     ELCONS     -1.0E14
    RHSZ       EAIRHT     4.0E9           EELEC       2.56E12
    RHSZ       EHISCAV    9.0E12          EHISOAV     3.0E12
    RHSZ       EHTCONST   1.20E13         ELCGCC      1.0E11
    RHSZ       ELEGO      3.80E11         ELNATGAS    4.0E12
    RHSZ       ELOCC      6.0E9           ELOLDCL     7.50E11
    RHSZ       ELOLDO     2.75E11         ELOSCAV     8.0E12
    RHSZ       ELOSOAV    3.0E12          ELRIVERW    3.12E13
    RHSZ       EMEDSCAV   1.0E13          ENEWCON     0.7E13
    RHSZ       EULPART    6.0E12          EULSO2      2.E10
    RHS2       ELNATGAS   -4.0E12
    RHS3       EZOTH2O    -1.0E15
ENDATA                      .0
```

Index

R

S

T

U

V

W

Z